The
MARINES
Have Landed

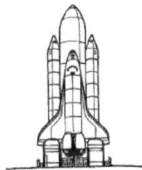

MSgt Andy Bufalo USMC (Ret)

ISBN 978-0-9817007-8-6

First Printing – March 2010
Printed in the United States of America

www.AllAmericanBooks.com

The Marines Have Landed

OTHER BOOKS BY ANDY BUFALO

HOLLYWOOD MARINES
Celebrities Who Served in the Corps

SWIFT, SILENT & SURROUNDED
Sea Stories and Politically Incorrect Common Sense

THE OLDER WE GET, THE BETTER WE WERE
MORE Sea Stories and Politically Incorrect Common Sense
Book II

NOT AS LEAN, NOT AS MEAN, STILL A MARINE!
Even MORE Sea Stories and Politically Incorrect Common Sense
Book III

EVERY DAY IS A HOLIDAY...
Every Meal is a Feast!
Yet Another Book of Sea Stories and Politically Incorrect Common Sense
Book IV

THE ONLY EASY DAY WAS YESTERDAY
Marines Fighting the War on Terrorism

HARD CORPS
The Legends of the Marine Corps

AMBASSADORS IN BLUE
In Every Clime and Place
Marine Security Guards Protecting Our Embassies Around the World

THE LORE OF THE CORPS
Quotations By, For & About Marines

"It is for me a touchstone of the Marine Corps' fatal glamour that there is no *ex*-Marine of my acquaintance, regardless of what direction he may have taken spiritually and politically after those callow gung-ho days, who does not view the training as a crucible out of which he emerged in some way more resilient, simply braver and better for the wear." – *Former Marine and Pulitzer Prize-winning author William Styron*

This book is for

**Jimmy McCain,
Son of Senator John McCain**

and

**Jimmy Webb,
Son of Senator Jim Webb**

Semper Fi

The Marines Have Landed takes up where *Hollywood Marines* left off. While the latter chronicles the exploits of famous Marines in the worlds of entertainment and sports, this companion book takes a look at Leathernecks who have made their mark in government and business. Many of the Marines profiled in this book will not be as familiar to the average reader as those found in the pages of *Hollywood Marines* because many of the politicians are regional or were from different eras, but that does not diminish what they accomplished. The amazing stories of Senator Paul Douglas, who joined the Marine Corps as a private during World War II at the age of fifty, and Mike Mansfield, whose headstone remembers him as a Marine PFC even though he was the longest serving majority leader in the history of the Senate, are for the ages. Not to be outdone, Marines who ventured into the private sector created or ran businesses like FedEx, Chrysler, Bank of America, Domino's Pizza and the *New York Times*. I hope this book serves as an inspiration to young Devil Dogs who are ready to serve our country in other ways once their duty is done because Marines have always been resilient and resourceful, and as the role models in these pages demonstrate, they can get virtually anything accomplished if they put their minds to it!

Semper Fi!

TABLE OF CONTENTS

The Marines Have Landed

BUSINESS & INDUSTRY

WALTER ANDERSON
Parade Magazine

Walter Anderson was the Chairman and CEO of Parade Publications from 2000 to 2009, and was editor of *Parade Magazine* for twenty years before becoming CEO.

Anderson was born in a tenement house in Mount Vernon, New York in 1944. His father was an illiterate and angry man and would beat young Walter every time he caught him trying to learn to read - however, Walter's mother believed the only hope for getting out of poverty was through learning. Anderson once said, "I could read, and be anyone and do anything, and I could imagine myself out of the slum. I *read* myself out of poverty long before I *worked* my way out of poverty."

The violence in his home pushed Anderson to drop out of high school, and he left to enlist in the Marine Corps. He served from 1961 to 1966, did a tour of duty in the Republic of Vietnam, and attained the rank of Sergeant.

During his enlistment Anderson earned his Graduate Equivalency Diploma (GED), and he has since been a national spokesman of for the GED. He also earned his college degree, graduating as valedictorian of his Mercy College class in 1982. Since that time he has written five books, including his memoir *Meant to Be*, which was published by HarperCollins in 2003. He is the recipient of the 1994 Horatio Alger Award, which honors individuals who have succeeded in the face of adversity.

LESLIE M. "BUD" BAKER, JR.
Wachovia Bank

Leslie M. "Bud" Baker, Jr. was the president and chief executive officer of Wachovia Corporation, one of the largest banks in the U.S. Most notably, he guided Wachovia through its 2001 merger with First Union, creating the nation's fourth largest bank with twelve million clients. He retired from Wachovia after thirty-four years with the company.

Baker served in the Marine Corps from 1964 through 1967, spent three years in Vietnam, and attained the rank of Captain. He graduated from the University of Richmond in 1964 with a degree in English literature, and received an MBA from the University of Virginia in 1969.

He is currently on the Board of Directors of Marsh & McLennan Companies, Inc., and is a director of the North Carolina Arboretum, Old Salem, Inc., and the James B. Hunt Institute for Education.

Bud Baker is also active with the Marine Corps Heritage Foundation, and lead a campaign to raise forty-five million dollars for the Foundation's Museum of the Marine Corps in Quantico, Virginia.

GLEN BELL
Taco Bell

Glen William Bell, Jr. (Sept 3, 1923 - Jan 16, 2010) was the businessman who founded the Taco Bell fast food chain.

Bell's first venture in the restaurant business was in 1948, when he opened Bell's Drive-In in San Bernardino, California. Bell founded his restaurant after he and San Bernardino High School classmate Neal Baker studied the success of the McDonald brothers and their namesake burger enterprise – which had also been founded in San Bernardino. The car culture was booming in 1948, and Bell was on the cusp of developing restaurants which offered revolutionary innovations to its customers including drive-ins, streamlined menus and quick service.

Bell's Drive-In first served a menu with hamburgers and hot dogs, however he soon decided to differentiate his menu by adding Mexican fare. He also understood the need to develop a convenient way to serve items such as tacos in a take-out environment, and began experimenting with a drive-thru concept.

Between 1954 and 1955 Bell and a business partner built three drive-thru taco stands in Southern California called 'Taco Tias,' but since his partner was not in favor of expanding into Los Angeles Bell sold his interest. In 1958

Bell and a new group of business partners opened 'El Tacos' in the Long Beach area, and although the chain expanded throughout California and was extremely profitable Bell soon sold his share of the business to his partners because he was ready to start his own venture.

In 1961 Bell, not content with perfecting the quick-service Mexican food concept, started 'Der Wienerschnitzel' together with an employee named John Galardi – who went on to build Der Wienerschnitzel into a chain of his own. Another employee, Ed Hackbarth, left to open a drive-in chain which would become known as Del Taco.

Taco Bell became a reality in 1962 when Bell opened his first restaurant in Downey, California. From there he expanded to a chain, sold the first Taco Bell franchise in 1964, and in 1978 sold his 868 Taco Bell restaurants to PepsiCo. Taco Bell is now owned by the world's largest restaurant company, Yum! Brands Inc.

A World War II veteran, Bell served in the Marine Corps and participated in the battles of Guadalcanal and Guam. He also saw post-war service in China before heading home to California.

After retiring from the restaurant business Bell built a 115-acre model produce farm and landscaped park named 'Bell Gardens' in Valley Center, California. It is open to the public and provides educational programs which stress the importance of agriculture and how to preserve our natural resources.

Glen Bell died in California on January 16, 2010 at the age of eighty-six.

MIKE ILITCH
Little Caesar's Pizza

Michael "Mike" Ilitch Sr. was born on July 20, 1929 in Detroit, Michigan. He is the owner of the Detroit Red Wings and Detroit Tigers, as well as founder and owner of the Little Caesars Pizza fast food franchise. A first generation American of Macedonian descent, he is married to Marian Bayoff Ilitch.

After graduating from Cooley High School, Ilitch entered the Marine Corps for four years. A sure-handed shortstop with a good bat, he had been scouted by the hometown Detroit Tigers and offered a $5,000 bonus to sign with the organization. Mike insisted on double that amount, and warned that if the Tigers wouldn't pay him a $10,000 bonus he would join the Marines. The Tigers balked, and in 1948 Mike signed up to serve his country.

As the Korean War escalated Ilitch was told he'd be sent to the front lines, but while on a military ship crossing the Pacific he was informed that when they stopped in Hawaii he would not be continuing with his buddies. "They took me off the ship so I could play for a (military) baseball team in Pearl Harbor," recalled Ilitch. "I finished my service there."

The Naval hospital at Pearl Harbor made an indelible impact on the young Marine. He recalled being horrified seeing wounded troops returning from Korea in wheelchairs and bandages. "It made a real impression," he said. "Could I have ended up that way if I had been sent to Korea? Absolutely!"

After his four-year tour of duty concluded Ilitch returned to Detroit, signed with the Tigers, and spent three years in the Detroit system - mostly with the Tampa Smokers of the Florida International League. His parents didn't approve of their son's career choice, but Mike had a dream and he worked hard to fulfill it. To make spending money while playing ball, Ilitch convinced a nightclub owner that he should be serving pizzas. The Detroit-area businessman hired Mike to make pizzas in a back room of the club, patrons loved the new treat, and Mike had stumbled onto an idea that was about to skyrocket. "I was fascinated by water and flour," he told the *New York Times*. "You knead it into dough, put it in the oven, and it comes out baked. Wow!"

A knee injury forced Ilitch out of his playing career, so instead he went into the pizza business. With the help of his wife Marian, Ilitch opened Little Caesars Pizza Treat in Garden City, Michigan. It was the first of what would become many thousands of restaurants through franchising. Today, the combined total revenues for their enterprises (as of 2007) exceed $1.8 billion.

Ilitch is an avid sports fan, and in 1982 he and Marian purchased the struggling Detroit Red Wings hockey franchise and turned it into a Stanley Cup champion. At the time of the purchase the team was known as the "Dead Wings," and interest in hockey in Detroit was at an all-time low. Since then the team has won eight divisional championships, four President's Trophies (for the season best record among all NHL teams), five Campbell Bowls and four Stanley Cups - in 1997, 1998, 2002 and 2008.

The Marines Have Landed

Then in 1992 Ilitch purchased the Detroit Tigers from fellow Marine and pizza magnate Tom Monaghan, founder of Domino's Pizza. Under his ownership, the Tigers logged losing records in twelve out of thirteen seasons before their turnaround in 2006, when the team made the playoffs for the first time in nineteen years. Ilitch also moved the team from Tiger Stadium into newly-built Comerica Park, and financed approximately fifty percent of the $350 million facility, with the taxpayers and federal grants covering the balance.

One of Ilitch's first philanthropic efforts was the Little Caesars Love Kitchen, established in 1985. The traveling restaurant was formed to feed the hungry and assist with food provisions during national disasters - most recently helping the flood victims and volunteers in North Dakota. The program has been recognized by former Presidents Bill Clinton, George H.W. Bush and Ronald Reagan, and has served more than two million individuals in the United States and Canada.

In 2006, inspired by a veteran returning to civilian life after losing both of his legs in war, Ilitch founded the Little Caesars Veterans Program to provide honorably discharged veterans with a business opportunity when they transition from service or seek a career change. Ilitch received the Secretary's Award from the Department of Veterans Affairs for this program in 2007. It is the highest honor given to a civilian by the Department. Today there are fifty Little Caesars Veteran franchisees who have applied more than $1.5 million in benefits.

The Ilitch family was presented the key to the City of Detroit by Mayor Kwame Kilpatrick on February 14, 2008. They are the fifth recipients of this award in the history of the city.

ALFRED LERNER
MBNA Bank

Alfred "Al" Lerner (May 8, 1933 - Oct 23, 2002) was a businessman who became chairman of MBNA Bank. Born in Brooklyn, New York, he was the son of Jewish-Russian immigrants.

Lerner attended Columbia University, and served in the Marine Corps from 1955 to 1957.

Lerner became chairman of MBNA Bank by investing eight hundred million dollars of his own money in the initial public offering (IPO) of the corporation. He also owned the Cleveland Browns of the National Football League, after purchasing the rights to the team in 1998. Prior to that, in 1995, he assisted his friend Art Modell - the former owner of the Browns - in moving Modell's NFL franchise rights from Cleveland to Baltimore. After Lerner's death his son Randy took over the Browns franchise, and Lerner's initials are stitched on the shoulders of the Browns' jerseys.

Lerner donated approximately twenty-five million dollars toward the construction of a new Columbia University student center in 1999, which was named Alfred Lerner Hall in his honor, and in 2007 Columbia announced it would honor Lerner's service in the Marine Corps with a plaque to be placed in the Hall. In addition the College of Business and

Economics at the University of Delaware is named after him, and while he was President of the Cleveland Clinic Foundation Lerner donated over one hundred million dollars to the hospital system.

Lerner was proud of being a Marine, and flew a Marine Corps flag atop the stadium the entire time he owned the Browns. His estate also donated ten million dollars toward the construction of the National Museum of the Marine Corps in Quantico.

Fellow Marine and Cleveland native Drew Carey also made a tribute to Lerner at the end of a season eight episode of *The Drew Carey Show* entitled "The Dawn Patrol."

ROBERT A. LUTZ
General Motors

Robert A. "Bob" Lutz was born on February 12, 1932 and is a vice chairman at General Motors Company.

Lutz was born in Zurich, Switzerland, the son of a bank director. He left Switzerland at the age of seven, and returned ten years later to attend school for a period in Lausanne. He received a Bachelor's degree in Production Management in 1961 followed by an MBA in 1962, both from UC Berkeley.

Lutz was CEO of Exide Technologies until the year it filed for bankruptcy and President of Chrysler Corporation, where he oversaw the development of the Dodge Viper, Plymouth Prowler and Chrysler LH platform automobiles.

He was also a Vice President at Ford Motor Company, where he led the creation of the Ford Sierra, initiated development of the original Ford Explorer, and spearheaded importation of models from Ford of Europe to the United States under the short-lived Merkur brand.

Prior to working at Ford he served as Executive Vice President of sales at BMW for three years, and he takes some credit in the development of the BMW 3-Series. He became one of only a few senior automotive executives with experience in both hemispheres, and with more than one major manufacturer, when he joined BMW after eight years with GM in Europe.

On February 9, 2009 GM announced that Lutz would step down from his position as Vice-Chairman of Global Product Development to take an advisory role. He said one reason for his decision was the increasing regulatory climate in Washington that would force him to design what Federal regulators wanted, rather than what customers wanted. Then, in a press conference on July 10, 2009, GM stated that Lutz would remain at GM as the vice chairman responsible for all creative elements of products and customer relationships.

Lutz is known as a collector of classic automobiles and military jets. Among other aircraft, he owns and pilots a Aero Vodochody L-39, an advanced Czechoslovakian jet fighter trainer. He also maintains a collection of motorcycles that include a Suzuki Hayabusa, a BMW K1200RS, a BMW K1200S, a BMW R1100S, and a BMW K-1.

An aviator in the Marine Corps, Lutz authored the management and leadership book *Guts*, which the dust jacket describes as "a maverick's primer on the business philosophy that revolutionized Chrysler."

Lutz has expressed skepticism on the issue of global warming, and on one occasion referred to it as "a total crock of shit." He maintains a blog called *Fastlane* that is hosted at GM Blogs, and is also a member of the board of the Marine Corps University Foundation and the Marine Military Academy.

HUGH MCCOLL
Bank of America

Hugh L. McColl Jr. was born 18 June 1935, and is a fourth-generation banker and former Chairman and CEO of Bank of America.

McColl was a driving force behind consolidating a series of progressively larger, mostly southern banks, thrifts and financial institutions into a super-regional banking force, "the first ocean-to-ocean bank in the nation's history" – a transformation that Tony Plath, director of banking studies at the University of North Carolina at Charlotte, called "the most significant banking story of the late Twentieth Century."

McColl was born in Bennettsville, South Carolina to cotton farmer and banker Hugh Leon McColl and artist Frances Pratt Carroll McColl. His great-grandfather, Duncan Donald McColl, had brought the first railroad as well as the first cotton mills to Marlboro County and had founded the Bank of Marlboro, later headed by McColl's grandfather and then his father.

McColl's father liquidated the Bank of Marlboro in 1939, and later bought a controlling interest in Marlboro Trust Co. Hugh went to work at age fourteen for the trust company and his father's cotton company, McColl Cotton Mills, keeping books, securing payments, learning double-entry accounting and driving across North and South Carolina to make deposits.

After graduating from the University of North Carolina at Chapel Hill, McColl joined the Marine Corps and served a two year tour of duty. Honorably discharged, he returned to North Carolina. Hugh McColl Sr. directed his son to banking, telling him he "didn't have the brains for farming."

On October 3, 1959 McColl married Jane Bratton Spratt McColl of York, South Carolina - the daughter of a banker, and sister of Congressman John Spratt (D-SC). McColl declined an offer from his father-in-law, John McKee Spratt, to work at the Bank of Fort Mill, a small family-owned bank, but let his father arrange an introduction at another bank. Subsequently, young McColl went to work as a management trainee for American Commercial Bank in Charlotte, North Carolina. In 1960, a year after McColl joined American Commercial, the bank joined with Greensboro's Security National Bank and became North Carolina National Bank. Vigorously competitive, McColl deployed a methodical, military approach to transforming the small regional bank, via incremental acquisitions and mergers, into Nations Bank and ultimately Bank of America.

McColl became President of NCNB in 1974 at age thirty-nine, and CEO in 1983 after the bank purchased the Lake City, Florida-based First National Bank of Lake City. NCNB then purchased First Republic Bank Corporation of Dallas, Texas from the FDIC and acquired over two hundred thrifts and community banks. In 1991, after acquiring Atlanta-based C&S/Sovran Corporation and becoming Nations Bank, the institution purchased in succession Maryland National Corporation, Chicago Research and Trading Group, Bank South, St. Louis-based Boatmen's Bancshares, Jacksonville, Florida based Barnett Bank, and Montgomery Securities

Then, in April of 1998 and under McColl's direction, Nations Bank and San Francisco-based Bank America merged, creating Bank of America.

After handing off day-to-day bank operations in 1999 and fully retiring from Bank of America in 2001, McColl partnered with other Charlotte banking executives to form McColl Partners, an investment banking firm which advises mid-sized companies on mergers and acquisitions.

McColl has supported a broad range of academic, civic and arts causes for Charlotte, the state of North Carolina and the Southeast by strongly encouraging Charlotte's urban redevelopment, playing a key role in attracting the Carolina Panthers NFL and Charlotte Hornets NBA franchises, supporting Habitat for Humanity, chairing the Forum for Corporate Responsibility, financing inner-city and minority-owned businesses, and encouraging light and high-speed rail.

Two books have been written about him - *McColl: The Man with America's Money* by Ross Yockey and *The Story of NationsBank: Changing the Face of American Banking* by Howard E. Jr. Covington and L. William Seidman.

McColl entered the South Carolina Business Hall of Fame in 1990, and in 1997 he was voted Tarheel of the Year. In 2005 he entered the North Carolina Business Hall of Fame, and in 2007 he entered the Junior Achievement U.S. Business Hall of Fame. McColl was named "Family Champion" by *Working Mother* magazine, earned the Pioneer Award from the Organization for a New Equality, and won the Applause Award from Women's Business Enterprise National Council.

The headquarters of the Kenan-Flagler Business School at UNC-Chapel Hill was also named the 'McColl Building' upon its completion in 1997 in recognition of McColl's efforts on behalf of his alma mater.

TOM MONAGHAN
Domino's Pizza

Thomas Stephen "Tom" Monaghan was born on March 25, 1937 in Ann Arbor, Michigan. He is an entrepreneur, conservative Catholic philanthropist, the founder of Domino's Pizza, and owned the Detroit Tigers from 1983 to 1992.

Monaghan sold Domino's in 1998, and has subsequently dedicated his time and considerable fortune to Catholic philanthropy and political causes. A champion of the pro-life and other conservative causes, Monaghan has spent hundreds of millions of dollars to further this agenda, garnering both praise and criticism.

Monaghan's father died when he was four years old and his mother had a difficult time raising him alone so in 1943, at the age of six, he ended up in an orphanage along with his younger brother until their mother collected them again in 1949. The orphanage, the St. Joseph Home for Children in Jackson, Michigan, was run by the Felician Sisters of Livonia. The nuns there inspired his devotion to the Catholic faith, and he later entered a minor seminary with the desire to eventually become a priest - but was expelled for a series of disciplinary infractions.

Then in 1956 Monaghan enlisted in the United States Marine Corps by mistake - he had actually meant to join the Army - and he received an honorable discharge in 1959.

Monaghan returned to Ann Arbor, Michigan in 1959 and enrolled in the University of Michigan with the intention of become an architect. While a student there, he and his brother James borrowed five hundred dollars to purchase a small pizza store called DomiNick's in Ypsilanti, Michigan. This business would, after enduring a lawsuit from Domino's Sugar, grow into Domino's Pizza. Tom, after opening a further three stores, traded his brother James a Volkswagen Beetle for his half of the business. Monaghan dropped sub sandwiches from the menu and focused on delivery to college campuses, inventing a new insulated pizza box to improve delivery. The new box, unlike its chipboard predecessors, could be stacked without squashing the pizzas inside. That permitted more pizzas per trip, and kept them warm until they arrived. By spreading his model to other college towns through a tightly-controlled franchising system, by the mid-1980s there were nearly three new Domino's franchises opening every day. He sold his controlling stake in Domino's Pizza in 1998 to Bain Capital, an investment firm based in Boston, for an estimated one billion dollars.

In 1983 Monaghan bought the Detroit Tigers, and they won the World Series a year later. He also became close to Major League Baseball commissioner Bowie Kuhn, who remained a good friend, business associate and participant in his many philanthropic works. Monaghan ultimately sold the Tigers to his competitor, fellow former Marine Mike Ilitch of Little Caesar's Pizza, in 1992.

Another of Monaghan's expensive passions has been automobiles, and for a time his collection included one of the world's six Bugatti Royales, for which he paid $8.1 million in 1986. He later sold the car for $8 million. In the early 1990s he also built a mission in a Honduras mountain town, and funded and supervised the construction of a new cathedral in Managua, Nicaragua after the old cathedral was destroyed in a 1972 earthquake.

The wealth Monaghan amassed from Domino's Pizza enabled a lavish lifestyle, however after reading a passage by C. S. Lewis on pride (from *Mere Christianity*) he divested himself of most of his more ostentatious possessions, including the Detroit Tigers, in 1992. He gave up his lavish office suite at Domino's headquarters, replete with leather-tiled floors and an array of expensive Wright furnishings, and turned it into a corporate reception room. He also ceased construction on a huge Frank Lloyd Wright-inspired mansion that was to be his home.

Monaghan is a Catholic with a particular interest in pro-life causes and the appointment of pro-life Justices to the U.S. Supreme Court. He has established or helped establish a number of Catholic organizations and educational establishments, and in 1987 received Holy Communion from Pope John Paul II in his private Papal chapel at the Vatican. His 1987 Vatican visit moved him so much he returned to the United States committed to promoting the Catholic faith. He soon established Ave Maria Radio, the Ave Maria List pro-life political action committee, and the Thomas More Law Center, a public interest law firm dedicated to promoting social conservative issues such as opposition to abortion, same sex marriage rights, and secularism. Monaghan publicly promotes daily attendance at Mass, daily recitation of the rosary and frequent sacramental confession. He has also committed to spending what remains of his $1 billion fortune on philanthropic endeavors.

When it came time for his story to be told, Tom Monaghan combined his passion for pizza and baseball by titling his 1986 autobiography *Pizza Tiger.*

JIMMY MURRAY
Ronald McDonald House

Jim Murray is the co-founder of the Ronald McDonald House and a former General Manager of the Philadelphia Eagles. He is a native of West Philadelphia, and is also president of Jim Murray Ltd, a sports promotion and marketing firm.

The son of Irish Catholic parents, Murray grew up in a row house in West Philadelphia and attended Our Mother of Sorrows Parish grade school and West Catholic High School. He graduated from Villanova University in 1960.

Murray began his career in sports administration with the Tidewater Tides of baseball's South Atlantic League. After a tour of active duty with the Marine Corps Reserve, he returned to baseball as assistant general manager of the Atlanta Crackers, an affiliate of the St. Louis Cardinals. Then in 1964 he left baseball to enter the restaurant business, but returned to Villanova as sports information director in 1966.

In 1969 Murray joined the Philadelphia Eagles' public relations staff and became the NFL team's administrative assistant two years later. Then in 1974, five years after joining the organization, he was named general manager of the Eagles. For more than nine years Murray served in that capacity, and took the franchise from the NFL's cellar to the

Super Bowl. In 1976 he and owner Leonard Tose hired Dick Vermeil as head coach, and from 1978 through 1981 the Eagles made the NFL playoffs every year. After the 1980 season the Eagles played the Raiders in Super Bowl XV, which is the only Super Bowl appearance in the franchise's history. Murray left the Eagles in 1983.

During his fourteen years with the Eagles Murray assumed leadership roles in a number of community projects. For one, he helped start the successful 'Eagles Fly for Leukemia' campaign. Most notably, he was the co-founder of the first Ronald McDonald House in Philadelphia along with Dr. Audrey Evans and persuaded many of his peers in the NFL to become involved in the unique Ronald McDonald House concept. Ronald McDonald Houses provide temporary homes, at little or no cost, for the families of children undergoing treatment for various illnesses at nearby hospitals. Started in Philadelphia in 1974, there are now over two hundred Ronald McDonald Houses worldwide.

Murray's numerous honors and awards include the first annual Leonard Tose Award in 2002, Citizen of the Year Award from the American Medical Association in 1999, the Distinguished Service Award from the American Legion in 1992, induction into the Philadelphia City All-Star Chapter of the Pennsylvania Sports Hall of Fame in 1992, President Ronald Reagan's Medal for Volunteers of America in 1987, the prestigious Bert Bell Man of the Year Award from the Bakers Club of Philadelphia in 1983, and the 2005 Award for Outstanding Catholic Leadership given by the Catholic Leadership Institute.

Jim Murray and his wife Dianne currently reside in Rosemont, Pennsylvania.

BOB PARSONS
GoDaddy.com

Robert "Bob" Parsons is a successful entrepreneur who was born in Baltimore, Maryland in 1950. He is the CEO and founder of the Go Daddy Group, Inc., a family of companies comprising three domain registrars, including flagship registrar GoDaddy.com, reseller registrar Wild West Domains, and Blue Razor Domains. His other affiliated companies include Domains by Proxy Inc., a domain privacy company, and Starfield Technologies, the business's technology development arm.

Parsons spent his childhood in Baltimore's inner city, and his family struggled financially. He has said about those days, "I've earned everything I've ever received. Very little was given to me. I've been working as long as I can remember. Whether it was delivering or selling newspapers, pumping gas, working in construction or in a factory, I've always been making my own money." By his own admission, he did not excel academically. Parsons often jokes that he was able to "enter the sixth grade with fourth grade skills," after somehow convincing his teacher to let him enter the sixth grade classroom even though he had failed to meet the requirements to pass fifth grade.

In 1968 Parsons joined the Marine Corps, and to this day insists that if he had not joined the military he would not have graduated from high school. He was assigned to the 26th Marine Regiment, which was attached to and operated as part of the 1st Marine Division, and in 1969 did a tour of duty in Vietnam as a rifleman in Delta Company of the 1st Battalion, 26th Marines in the Quang Nam Province.

"I can remember arriving on that hill in the middle of nowhere," Parsons commented. "The night prior, the squad I was newly assigned to had been ambushed and most of them were killed. It wasn't that moment I got afraid. Later that night I sat on the wall of an old French fort and looked into this valley and thought 'this is it... I'm going to die here,' and I accepted that, and from then on I was okay."

Parsons then did something that he uses to this day. "From the next day on my only goal, and I mean my *only* goal, was to simply make it to mail call. I figured that I would take small steps, tie them together, and it would get me to the finish line and that thought is what I use today to make GoDaddy successful - simply get to the next day."

He was later wounded there, medically evacuated, and spent two months at a naval hospital recovering from his wounds. As a result of his service and injury he earned the Combat Action Ribbon, the Vietnam Gallantry Cross, and the Purple Heart.

After Vietnam Parsons obtained an accounting degree, graduating magna cum laude in 1975 from the University of Baltimore. Prior to graduating he stumbled into a bookstore on the Stanford campus in San Francisco, found a book on computer programming, and read it on the way home. He started experimenting by writing programs with no instruction whatsoever - and a computer programmer was born. He then began his long-term career in software as a self-taught programmer. "My hobby of writing programs would soon become my career," Parsons said. "My first computer was a Radio Shack, then an Apple, and finally I borrowed five thousand dollars and bought the first IBM personal computers to come out."

In 1984 he founded Parsons Technology in a basement in Cedar Rapids, Iowa and began selling MoneyCounts, a home accounting program - and by late 1987 was able to quit his job and focus completely on selling and programming. Parsons Technology eventually grew to be a one-thousand-employee privately held company, and in 1994 he sold the business to Intuit for sixty-four million dollars.

"I saw the writing on the wall as this industry was shrinking," comments Parsons. "So, I sold the company to Intuit, retired, and moved to Arizona. The problem was I could not sit still, so in 1997, just as a hobby or for something to do, I started GoDaddy.com."

"There is no doubt in my mind that the lessons I took away from the Corps have made me succeed in my life," said Parsons. "The Marine Corps taught me and gave me the sense of doing things right, and more importantly I got confidence and that part of it is so important. I decided to offer people a great product or service for a reasonable and fair price. It's that simple - make a small amount of money on a whole lot of people, treat them right, and it will all fall into place."

One last thing - make sure that you take a look at the annual Marine Corps birthday card he puts out on his website!

CHARLES PHILLIPS
Oracle Corporation

Charles E. Phillips was born in Little Rock, Arkansas in 1959 and is President of Oracle Corporation as well as a member of the company's Board of Directors. His responsibilities encompass global field operations.

Prior to joining Oracle, Phillips was a Managing Director with Morgan Stanley where he currently serves as a director. He is also a director at Viacom Corporation, the American Museum of Natural History, and Jazz at Lincoln Center. Prior to his career on Wall Street, Phillips was a Captain in the United States Marine Corps and served with the 2nd Battalion, 10th Marines Regiment.

He holds a BS in Computer Science from the United States Air Force Academy, an MBA from Hampton University, and a JD from New York Law School. In February of 2009 Phillips was appointed as a member of the President's Economic Recovery Advisory Board to provide President Obama and his administration with advice and counsel in addressing the late-2000s recession.

In January of 2010 Phillips appeared in the headlines as a result of a billboard and an online campaign by YaVaughnie Wilkins, his alleged ex-mistress. While the billboards were taken down after a number of days, the mainstream media had already run the story. As a result, a representative of Mr. Phillips issued a statement quoting him as saying, "I had an eight and a half year serious relationship with YaVaughnie Wilkins. My divorce proceedings began in 2008, the relationship with Ms. Wilkins has since ended, and we both wish each other well." This coincided with an announcement from the European Union regulatory commission that the long-awaited purchase of Sun Microsystems by Oracle Corporation would be unconditionally approved.

LAWRENCE G. RAWL
Exxon Corporation

Lawrence G. Rawl (May 4, 1928 - February 14, 2005) was the Chairman and CEO of Exxon from 1985 to 1993.

Rawl was born in New Jersey in 1928, and toward the end of World War II he enlisted and served in the U.S. Marine Corps. In 1952 he graduated with a Bachelor of Science degree in petroleum engineering from the University of Oklahoma, and joined Humble Oil and Refining as drilling engineer.

By 1980 Rawl was named a senior vice president and director of Exxon Corporation, in 1985 he was named president of the corporation, and in 1987 he became chairman and CEO. During his tenure as head of Exxon he moved the corporate headquarters from New York to Irving, Texas, increased reserves, and expanded the chemical operations of the corporation.

Rawl was at the helm of the company when the *Exxon Valdez* went aground in 1989. He faced criticism for his response to the resultant oil spill - his slow public response and demeanor in interviews were noted, and became the focus of criticism of the company.

Lawrence Rawl retired from Exxon in 1993 at the mandatory retirement age of sixty-five after forty-one years with the company. He died in Fort Worth, Texas in February of 2005 at the age seventy-six.

41

FREDERICK W. SMITH
FedEx

Frederick Wallace "Fred" Smith was born in Marks, Mississippi on August 11, 1944 and is the founder, chairman, president, and CEO of FedEx. Originally known as Federal Express, it is the first overnight express delivery company in the world, and the largest in the United States.

Smith is the son of James Frederick Smith, the founder of the Toddle House restaurant chain and the Smith Motor Coach Company. In 1931 the Greyhound Corporation bought a controlling interest in the latter, and renamed it the Dixie Greyhound Lines. Fred's father died while he was only four, and the boy was raised by his mother and uncles.

In 1962 Smith entered Yale University, and while there wrote a paper for an economics class outlining overnight delivery service in a computer information age. Folklore suggests he received a C for this paper, although in a later interview he claims that when asked he told a reporter, "I don't know what grade, probably made my usual C," while other tales suggest his professor told him that in order for him to get a C, the idea had to be feasible. The paper became the idea for FedEx, and for years the sample package displayed in the company's print advertisements featured a return address at Yale.

Smith became a member of Delta Kappa Epsilon fraternity and the secret society Skull and Bones, and received his Bachelor's degree in economics in 1966. During his college years he was a friend of George W. Bush, and also befriended and shared an enthusiasm for aviation with John Kerry.

After graduation Smith joined the Marine Corps and served for four years, from 1966 to 1969, as a platoon leader and forward air controller (FAC) flying in the back seat of an OV-10. While technically a "Ground Officer" for his entire time in the Corps, he was specially trained to fly with pilots to observe and 'control' ground action. He himself never went through flight training, and was not a Naval aviator.

As a Marine, Smith had the opportunity to observe the military's logistics system first hand. He served two tours of duty in Vietnam, flew over two hundred combat missions, and was honorably discharged in 1969 with the rank of Captain - having received the Silver Star, the Bronze Star, and two Purple Hearts. All the while Smith carefully observed the procurement and delivery procedures and fine-tuned his dream for an overnight delivery service.

In 1970 Smith purchased the controlling interest in an aircraft maintenance company, Ark Aviation Sales, and by 1971 turned its focus to trading used jets. The following year he founded Federal Express with his four million dollar inheritance, and raised an additional ninety-one million in venture capital. By 1973 the company was offering service to twenty-five cities, beginning with small packages and documents and a fleet of fourteen Falcon 20 jets. His focus was on developing an integrated air-ground system, which had never been done before. Smith developed FedEx on the business idea of a shipment version of a bank clearing house where one bank clearing house was located in the middle of the representative banks and all their representatives would be sent to the central location to exchange materials.

In the ensuing years Smith has served on the boards of several large public companies, the St. Jude Children's Research Hospital, and the Mayo Foundation. He was also approached by Senator Bob Dole, who asked Smith for support in opening corporate doors for a new World War II memorial - and subsequently appointed him co-chairman of the U.S. World War II Memorial Project.

In addition to FedEx, Smith is co-owner of the NFL's Washington Redskins. His son Arthur, who played football at the University of North Carolina, is now a coach for the team. This partnership resulted in a FedEx sponsorship of the Joe Gibbs NASCAR racing team. Smith also owns or co-owns several entertainment companies, including Dream Image Productions and Alcon Films.

As a Fraternity Brother of George W. Bush while at Yale, after the 2000 election there was some speculation that Smith might be appointed to the Bush Cabinet as Defense Secretary - but Donald Rumsfeld was named instead. Although Smith was friends with both 2004 major candidates, John Kerry and George W. Bush, he chose to endorse Bush for re-election.

Fred Smith was inducted into the Junior Achievement U.S. Business Hall of Fame in 1998. He was awarded "CEO of the Year 2004" by *Chief Executive Magazine,* presented the 2008 Kellogg Award for Distinguished Leadership by the Kellogg School of Management, and awarded the 2008 Bower Award for Business Leadership from The Franklin Institute. He is also a member of the Aviation Hall of Fame.

EUGENE STONER
Father of the M16

Eugene Morrison Stoner (Nov 22, 1922 - April 24, 1997) is the man most associated with the design of the AR-15, which was adopted by the military as the M16. He is regarded by most historians, along with John Browning and John Garand, as one of the most successful military firearms designers of the 20th century.

Stoner attended high school in Long Beach, California and afterwards worked for the Vega Aircraft Company installing armament. During World War II he enlisted in the Marine Corps for Aviation Ordnance, and served in the South Pacific and northern China.

In late 1945 he began working in the machine shop for Whittaker, an aircraft equipment company, and ultimately became a Design Engineer. Then in 1954 he came to work as chief engineer for ArmaLite, a division of Fairchild Engine & Airplane Corporation. While at ArmaLite he designed a series of prototype small arms, including the AR-3, AR-9, AR-11, and AR-12, none of which saw significant production. The only real success during this period was the AR-7 survival rifle, which was adopted by the United States Air Force.

In 1955 Stoner completed initial design work on the revolutionary AR-10, a lightweight (7.25 lbs.) selective-fire infantry rifle in 7.62 x 51 mm NATO caliber, and it was submitted to the U.S. Army's Springfield Armory for rifle evaluation trials late in 1956. In comparison with competing rifle designs previously submitted for evaluation the AR-10 was smaller, easier to fire in automatic, and much lighter - however it arrived very late in the testing cycle, and the Army rejected it in favor of the more conventional T44, which would become the M14. The AR-10's design was later licensed to the Dutch firm of Artillerie Inrichtingen, which produced the rifle until 1960 for sale to various foreign military forces.

At the request of the U.S. military Stoner's chief assistants, Robert Fremont and Jim Sullivan, designed the AR-15 from the basic AR-10 design by scaling it down to fire the small-caliber .223 Remington cartridge. The AR-15 was later adopted by United States military forces as the M16 rifle.

After ArmaLite sold the rights to the AR-15 to Colt, Stoner turned his attention to the AR-16 design. This was another advanced 7.62 mm rifle, but it used a more conventional piston and a number of stamped parts to reduce cost. This weapon saw only prototype development but adaptation to .223 resulted in the somewhat successful and often imitated ArmaLite AR-18.

Stoner left ArmaLite in 1961 to serve as a consultant for Colt, and eventually accepted a position with Cadillac Gage where he designed the Stoner 62 Weapons System. This was a modular weapons system which could be configured as a standard automatic rifle, a light machine gun, a medium machine gun, or a solenoid-fired fixed machine gun. The Stoner Weapons System used a piston-operated gas impingement system, though Stoner himself believed direct gas operation was the ideal method for firearms. Once again,

Robert Fremont and Jim Sullivan would take a Stoner design and redesign it for the .223 Remington cartridge, to create the Stoner 63.

Stoner co-founded ARES Incorporated of Port Clinton, Ohio in 1972, but left the company in 1989 after designing the Ares Light Machine Gun, an evolved version of the Stoner 63 sometimes known as the Stoner 86. At Ares, he also developed the Future Assault Rifle Concept (FARC).

In 1990 he joined Knight's Armament Company to create the Stoner Rifle-25 (SR-25), which currently sees military service as the United States Navy Mark 11 Mod 0 Sniper Weapon System. While at KAC he also worked on yet another version of the Stoner Weapons System called the Stoner 96, and among his last designs was the SR-50 rifle.

Eugene Stoner died on April 24, 1997 in Palm City, Florida at the age of seventy-four.

ARTHUR OCHS SULZBERGER
The New York Times

Arthur Ochs "Punch" Sulzberger was born in New York City on February 5, 1926 to a prominent media and publishing family. He is himself an American publisher and businessman, and succeeded both his father, Arthur Hays Sulzberger, and his maternal grandfather, as publisher and chairman of the New York Times in 1963. He eventually passed the position to his son Arthur Ochs Sulzberger Jr. in 1992.

Sulzberger graduated from the Loomis Institute and then enlisted into the United States Marine Corps during World War II, serving from 1944 to 1946 in the Pacific Theater. He earned a B.A. degree in English and History in 1951 at Columbia University, and upon graduation was recalled to active duty by the Marine Corps Reserve because of the Korean War. Following completion of officer training he saw duty in Korea, and then in Washington, D.C., before being returned to reserve status.

Sulzberger became publisher of *The Times* in 1963 after the death of his brother-in-law, Orvil Dryfoos. During the 1960s he built a large news-gathering staff at *The Times*, and was publisher when the newspaper won a Pulitzer Prize in 1972 for printing The Pentagon Papers. Renowned first amendment attorney Floyd Abrams commented about the decision, saying, "Eventually Sulzberger, then in London, and rejecting the views of some of his colleagues in senior management as well as the dire warnings of his outside counsel, made the call to accept the risks of publication rather than those of silence. On Sunday, June 13, (1971), the *Times* published the first in a series of seven articles about the Pentagon Papers. In retrospect the decision may seem obvious, but it was by no means an easy one at the time, and it remains one for which Sulzberger deserves enormous credit."

Although Sulzberger relinquished the position of publisher to his son Arthur Ochs Sulzberger Jr. in 1992, he remained Chairman of The New York Times Company until October of 1997.

In 2005, the Newspaper Association of America (NAA) honored Punch Sulzberger with the Katharine Graham Lifetime Achievement Award.

The Marines Have Landed

POLITICS

HENRY BELLMON
Governor of and Senator from Oklahoma

Henry Louis Bellmon (Sept 3, 1921 - Sept 29, 2009) was a Republican politician from Oklahoma, a member of the Oklahoma Legislature, the 18th and 23rd Governor of Oklahoma (and the first Republican to hold that office), and a two-term United States Senator.

Bellmon was born in Tonkawa, Oklahoma, attended Billings High School in Billings, Oklahoma and graduated from Oklahoma A&M (now Oklahoma State University) in 1942 with a Bachelors Degree in agriculture. He was a Lieutenant in the Marine Corps from 1942 to 1946, served as a tank platoon leader in the Pacific Theater during World War II, and took part in four amphibious landings - including Iwo Jima. For his service he was awarded the Legion of Merit and a Silver Star, and after the war he returned to farming and took up politics.

Bellmon served a single term in the Oklahoma House of Representatives from 1946 to 1948, and while in the legislature in January of 1947 he married Shirley Osborn, to whom he remained married until her death in 2000. In 1960 he served as the State Republican Party Chairman, and in 1962 he was elected Oklahoma's first Republican Governor since statehood in 1907 and served his first term from 1963 to 1967. While Governor he served as the chairman of the

Interstate Oil Compact Commission, and also as a member of the executive committee of the National Governor's Association. Under Oklahoma law at the time there was a term limit, so he was not able to run for a second term.

In 1968 Bellmon was serving as the national chairman for Richard Nixon's presidential election campaign, but then decided to run for the U.S. Senate and won, unseating U.S. Senator A.S. Mike Monroney. In the Democratic landslide of 1974 he managed to be reelected over Congressman Ed Edmondson by a very narrow margin, but decided not run for a third term in 1980.

During his service in the Senate Bellmon sometimes took moderate positions which put him at odds with the largely conservative Oklahoma Republican Party. He supported Gerald Ford over Ronald Reagan in the 1976 presidential election even though the state delegation was committed to Reagan, he opposed a constitutional amendment to prohibit forced busing for the purpose of racially desegregating public schools, and he supported the Panama Canal Treaty.

In 1986 Oklahoma Republican leaders asked Bellmon if he would consider running for Governor again, since the term limit provision had been removed. He agreed, and that November Oklahoma voters returned him to the Governor's Mansion for a second term. During that term Bellmon worked with Democrats in the Oklahoma legislature to pass an educational reform package over the opposition of most Republicans, and when he chose not to seek reelection in 1990 the Republican candidate who replaced him, Bill Price, promised to repeal the bill but lost to Democrat David Walters, whom Bellmon had defeated four years earlier.

Henry Bellmon died on September 29, 2009 in Enid, Oklahoma at the age of eighty-eight after a long battle with Parkinson's disease.

DANIEL B. BREWSTER
Senator from Maryland

Daniel Baugh Brewster (Nov 23, 1923 - Aug 19, 2007) was a Democratic member of the United States Senate who represented the State of Maryland from 1963 until 1969. He was also a member of the Maryland House of Delegates from 1950 to 1958, and a Representative from the 2nd Congressional District of Maryland in the United States House of Representatives from 1959 to 1963.

Brewster was born on November 23, 1923 in Baltimore County, Maryland in the Green Spring Valley Region, the son and one of six children of Daniel Baugh Brewster Sr. and Ottolie Y. Wickes. His political lineage was quite extensive. He was a direct lineal descendant of Elder William Brewster, the Pilgrim colonist leader and spiritual elder of the Plymouth Colony who was a passenger on the *Mayflower*. He was also a great-grandson of Benjamin Harris Brewster, who served as United States Attorney General from 1881 to 1885, a great-great-great-great-grandson of Benjamin Franklin, and a relative of George Mifflin Dallas, a U.S. Senator from Pennsylvania, and the eleventh Vice President of the United States under President James K. Polk.

Brewster was educated at the Gilman School in Baltimore City and St. Paul's School in Concord, New Hampshire. He then attended college at Princeton and Johns Hopkins

Universities before the U.S. entry into World War II. He was enrolled at Princeton at 1942, but left to volunteer for the Marines at age nineteen and was commissioned from the ranks in 1943. He fought on Guam in 1944 and Okinawa in 1945, was wounded seven times, received a Bronze Star with a Gold Star in lieu of a second award, and left active duty in 1946. He remained in the Reserves until 1972, and attained the rank of Colonel.

After the war Brewster completed his undergraduate education at Johns Hopkins, and then enrolled in the University of Maryland Law School. He was admitted to the Bar in 1949, and commenced law practice in Towson, Maryland soon after.

In 1958 Brewster was elected to the House of Representatives from the 2nd District of Maryland, and in 1962 he ran for the United States Senate seat vacated by retiring Republican Senator John Marshall Butler and defeated Congressman Ted Miller to become the first Democrat elected to the Senate from Maryland since 1946. He served in the Senate from 1963 to 1969, when he was defeated in the 1968 election by Charles Mathias, Jr.

In 1964 Brewster ran in the Democratic presidential primaries against segregationist George Wallace. He focused on issues ranging from the presence of communist troops in Cuba in 1963 to proposed cuts in weekend postal service in 1964. His concern with mail practices continued in 1965 when he criticized the "mail cover" practice which permitted holding up mail to and from persons under investigation. Stressing the importance of the right of privacy, Brewster urged Postmaster General Larry O'Brien to ban the practice except in the case of treason or national security.

In a November 1966 letter to the *New York Times*, Brewster declared his support for advertising or "junk" mail, which he claimed accounted for thirty-five billion in sales. Pointing out that eighty percent of the mail was for business purposes, Brewster expressed concern over possible unemployment in private business and the postal service if junk mail was eliminated. In 1967, he voted for a "junk mail amendment" which would delay price increases and limit third class mail rates. Brewster also played a strong supporting role in national Democratic politics.

In 1969 Brewster was indicted on ten criminal counts of solicitation and acceptance of bribes while a United States Senator in his role as a member of the Committee on Post Office and Civil Service, as well as two counts of accepting illegal gratuities. This stemmed from a campaign contribution by Spiegel, a mail-order firm. He contended that he had done nothing wrong.

At trial the judge dismissed five of the charges, saying that his actions were protected under the Speech and Debate Clause of the U.S. Constitution. The prosecution appealed directly to the U.S. Supreme Court, which heard the case in 1972 and ruled that the taking of illegal bribes was not protected as part of the "performance of a legislative function."

The charges were reinstated, Brewster stood trial, and he was convicted - but in 1974 his conviction was overturned on appeal due to the trial judge's improper instructions to the jury. Then in 1975 he pleaded no contest to a misdemeanor charge of accepting an illegal gratuity and was fined and allowed to keep his law license, with the government dropping the other charges.

After leaving the Senate, Daniel Brewster took up farming in Glyndon, Maryland. He died of liver cancer on August 19, 2007 at the age of eighty-three.

DALE BUMPERS

Governor of and Senator from Arkansas

Dale Leon Bumpers served as the 38th Governor of Arkansas from 1971 to 1975, and as a United States Senator from 1975 until his retirement in January of 1999. A Democrat, he is currently counsel at the Washington, D.C. law office of Arent Fox LLP, where his clients include Riceland Foods and the University of Arkansas for Medical Sciences.

Bumpers was born on August 12, 1925 in Charleston, Arkansas, which is near Fort Smith. He attended public schools, the University of Arkansas at Fayetteville, and served in the Marine Corps from 1943 to 1946 during the Second World War. He then graduated from Northwestern University Law School in Evanston, Illinois in 1951, was admitted to the Arkansas bar in 1952, and started practicing law in his hometown that same year. He also served as Charleston City attorney from 1952 to 1970, and as special justice of the Arkansas Supreme Court in 1968.

Bumpers was virtually unknown when he announced his campaign for Governor in 1970, but despite his lack of name recognition his oratorical skills, personal charm, and outsider image put him in a runoff election for the Democratic nomination with former Governor Orval Faubus. Bumpers easily defeated Faubus, and then unseated incumbent moderate Republican Governor Winthrop Rockefeller in the general election. It was a heavily Democratic year nationally, and the tide benefited Bumpers. He, like Jimmy Carter of Georgia and John C. West of South Carolina, was often described as a new kind of Southern Democrat who would bring reform to his state and the Democratic Party. His victory over Rockefeller ushered in a new era of youthful reform-minded Governors, including two of his successors, David Pryor (later a three-term U. S. Senator) and future president Bill Clinton.

In the 1972 Democratic primary Bumpers easily defeated two opponents, including highly regarded State Senator Q. Byrum Hurst of Hot Springs. Then in the general election he swamped Republican Len E. Blaylock of Perry, even as Richard M. Nixon was handily winning Arkansas in the presidential race.

Bumpers was elected to the United States Senate in 1974, unseating incumbent James William Fulbright in the party primary, and then overwhelming Republican John Harris Jones on Election Day. It was to be the first of four terms.

Bumpers chaired the Senate Committee on Small Business and Entrepreneurship from 1987 until 1995, and when the Republican Party took control of the Senate for a dozen years following the 1994 elections he served as ranking minority member of the Senate Committee on Energy and Natural Resources from 1997 until his retirement. In the Senate, Bumpers was known for his oratorical skills and his prodigious respect for the Constitution - in fact, he never supported any constitutional amendment.

Despite support from many colleagues, including the ultimate 1988 Democratic candidate Senator Paul Simon of Illinois, Bumpers decided to not seek the Democratic Presidential nomination in 1984 and 1988. He was initially named as one of Walter Mondale's top potential choices for his vice presidential running mate in '84, but took his name out of the running early in the process. Bumpers stated as his main reason for not running the fear of, "a total disruption of the closeness my family has cherished." Many observers felt Bumpers perhaps lacked the obsessive ambition required of a presidential candidate, especially one who would have started out the process with low name recognition. Another factor often mentioned was Bumpers key vote in killing labor law reform in 1978, which angered organized labor and had clearly not been forgotten by labor leaders nearly a decade later.

Bumpers, a close friend of President Bill Clinton, gave an impassioned closing argument in his defense during Clinton's impeachment trial. Later, because Bumpers and his wife Betty were both known for their dedication to the cause of childhood immunization, the Dale and Betty Bumpers Vaccine Research Center (VRC) at the National Institutes of Health was established by former President Clinton to facilitate research in vaccine development.

In 1995, the University of Arkansas-Fayetteville founded the Dale Bumpers College of Agricultural, Food and Life Sciences in his honor.

CONRAD BURNS
Senator from Montana

Conrad Ray Burns, a former United States Senator from Montana, is only the second Republican to represent that state in the Senate since the passage of the Seventeenth Amendment in 1913. He is also the longest serving Republican United States Senator in Montana history, serving from January 1989 to January 2007.

Burns was born on a farm near Gallatin, Missouri to Russell and Mary Frances (Knight) Burns on January 25, 1935. He graduated from Gallatin High School in 1952, and then enrolled in the College of Agriculture at the University of Missouri. Two years later, in 1955, Burns left without graduating and enlisted in the Marine Corps. He served in Japan and Korea as a small-arms instructor.

After his military service ended in 1957 Burns began working for Trans World Airlines and Ozark Airlines, and in 1962 he traveled the state as a field representative for *Polled Hereford World* magazine in Billings, Montana. Then in 1968 he turned down a transfer to Iowa, and became a cattle auctioneer for the Billings Livestock Commission. He also began reporting on agricultural market news, started a radio show, and later worked as a farm reporter for KULR-TV.

In 1975 Burns founded the Northern Agricultural Network with four radio stations. The NAN had grown to serve thirty-one radio stations and six television stations by 1986 when he sold it to enter politics. Burns, angry at a local politician, ran and won a seat on the Yellowstone County Commission, where he served for two years.

Though a political novice, Burns had a great deal of name recognition from his previous jobs, and the Republican Party recruited him to run against Democrat and incumbent Senator John Melcher, a popular veterinarian. Burns portrayed Melcher as "a liberal who is soft on drugs, soft on defense, and very high on social programs." At the time a strong supporter of term limits, Burns said Melcher had been in Washington, D.C. for too long. Melcher was also hurt by public opposition to policies in Yellowstone National Park regarding naturally occurring fires, and by President Ronald Reagan's pocket veto of Melcher's bill, which would have made 1.4 million acres of Montana forest off-limits to logging and mineral development. Burns defeated Melcher in a close race, 51-48 percent, and was no doubt helped by the success in Montana of Vice President George H. W. Bush, the Republican presidential nominee.

Burns faced Brian Schweitzer, a rancher from Whitefish, Montana, when he ran for reelection in 2000. While Burns attempted to link Schweitzer with presidential candidate Al Gore, whom Schweitzer had never met, Schweitzer effectively portrayed himself as being nonpolitical. Schweitzer primarily challenged Burns on the issue of prescription drugs, and organized busloads of senior citizens to take trips to Canada and Mexico for cheaper medicine. In turn, Burns charged that Schweitzer favored "Canadian-style government controls."

Burns spent twice as much money as Schweitzer on the election and only defeated him by a slim margin, 51-47 percent, while the state voted 58-33 percent for Republican presidential nominee George W. Bush. Schweitzer went on to become Governor in 2004.

In December of 2003 Burns and Senator Ron Wyden, an Oregon Democrat, were pleased when their legislation to combat spam, the CAN-SPAM Act, was signed into law. Burns said, "Senator Wyden and I have worked during this time to come up with common-sense legislation to deal with spam, and I think we've been successful."

Because of his narrow win in 2000 and the Democratic takeover over of Montana's state government in 2004, polls in 2006 put Burns' support in the state at around forty percent, making him one of the most vulnerable Senators facing re-election in 2006.

Burns' Democratic opponent was Montana State Senate President and organic farmer Jon Tester. Since at least August, the Montana Democratic Party had paid staffer Kevin O'Brien to follow Burns around the nation, continuously filming the Senator at all his public events, including Senate committee hearings and various campaign appearances. The content of these tapes were used for a variety of "gotcha" press releases and even a YouTube music video.

In an October election debate with Tester regarding the Iraq War, Burns said that Tester "says our president doesn't have a plan. I think he's got one. He's not going to tell everyone in the world," and later told Tester, "We're not going to tell you what our plan is Jon, because you're just going to go out and blow it." On November 9, 2006 Burns conceded the election to Tester.

After leaving office Burns joined the lobbying firm of GAGE, founded by his former chief of staff Leo Giacometto. He also founded a communications company called Rural Solutions Corporation to expand broadband communications in rural areas, and maintains his auction business where he arranges speaking engagements.

On December 9, 2009 Conrad Burns was hospitalized after suffering an atrial fibrillation and having a stroke at his Arlington, Virginia office. He was taken to Virginia Hospital Center before being transferred to the intensive-care unit of a Washington, D.C. hospital.

JAMES CARVILLE
Political Strategist

James Carville, known variously as the "Ragin' Cajun" or "Corporal Cue Ball," is a political consultant, actor, attorney, media personality, and prominent liberal pundit. Carville first gained national attention for his work as the lead strategist in the successful presidential campaign of then-Arkansas Governor Bill Clinton. Carville was also a co-host of CNN's *Crossfire* until its final broadcast in June of 2005. Since its cancellation he has appeared on CNN's news program *The Situation Room,* and as of 2009 hosts a weekly program on XM Radio titled *60/20 Sports.*

Carville, the oldest of eight children, was born Chester James Carville, Jr. at Fort Benning, Georgia on October 25, 1944, the son of Lucille (née Norman), a former school teacher who sold World Book Encyclopedias door-to-door, and Chester James Carville, a postmaster as well as owner of a general store. He is of Irish and Cajun ancestry, was raised in Carville, Louisiana, and attended Ascension Catholic High School in Donaldsonville, Louisiana.

Carville graduated from Louisiana State University with undergraduate and law degrees, and before entering politics worked as a litigator at a Baton Rouge law firm from 1973-1979. He also spent two years serving in the Marine Corps, and worked as a high school teacher.

Prior to the Clinton campaign Carville and consulting partner Paul Begala gained other well-known political victories, including the gubernatorial victories of Robert Casey of Pennsylvania in 1986 and Zell Miller of Georgia in 1990, but it was in 1991 that they rose to national attention by leading appointed incumbent Senator Harris Wofford of Pennsylvania back from a forty-point poll deficit over the White House's hand-picked candidate, Dick Thornburgh. Also noteworthy is Wofford's campaign was where the "it's the economy, stupid" strategy used by Bill Clinton in 1992 was first implemented.

In 1992 Carville helped lead Clinton to a win against George H. W. Bush in the Presidential election, and in 1993 he was honored as Campaign Manager of the Year by the American Association of Political Consultants. His role on the Clinton campaign was documented in the feature-length Academy Award-nominated film, *The War Room*. One of the formulations he used in that campaign has entered the language, derived from a list he posted in the war room to help focus himself and his staff, with these three points: Change vs. more of the same. The economy, stupid. Don't forget health care.

In 2006, Carville switched gears from politics to sports and became a host on a sports show called *60/20 Sports* on XM Satellite Radio with Luke Russert, the son of late NBC journalist Tim Russert. The show is an in-depth look at the culture of sports based on the ages of the two hosts (sixty and twenty).

On March 4, 2009 *Politico* reported that Carville, Paul Begala, and Rahm Emanuel were the architects of the Democratic Party's strategy to cast conservative talk radio host Rush Limbaugh as the face of the Republican Party. Carville was particularly critical of Limbaugh for saying he wanted Barack Obama to "fail." It was later reported that Carville had voiced the opinion, during the presidency of George W. Bush, that, "I don't care if people like him or not, just so they don't vote for him and his party. That is all I care about. I hope he doesn't succeed, but I am a partisan Democrat. But the average person wants him to succeed. It is his country, his life or their lives. So he has that going for him." Carville made the remarks on September 11, 2001, shortly before the terrorist attacks on the United States. Upon hearing news of the attacks, Carville asked reporters to "disregard" his prior comments.

As an advisor to Hillary Rodham Clinton's 2008 presidential campaign, Carville told *The New York Times* on March 22, 2008 that New Mexico Governor Bill Richardson, who had just endorsed Senator Barack Obama for the Democratic nomination, was comparable to Judas Iscariot. It was "an act of betrayal," said Carville. "Mr. Richardson's endorsement came right around the anniversary of the day when Judas sold out for thirty pieces of silver, so I think the timing is appropriate, if ironic," referring to Holy Week. Governor Richardson had served in President Bill Clinton's administration as both United States Ambassador to the United Nations and Secretary of Energy, and Carville believed that Richardson owed an endorsement to Senator Clinton in exchange for being offered those posts by her husband. Carville also claimed that Richardson assured many in the Clinton campaign that he would at least remain neutral and abstain from taking sides. Richardson refuted Carville's account, arguing that he had not made any promises to remain neutral. Richardson claims that his decision to endorse Obama was "clinched" by his speech on race relations following the swirl of controversy surrounding Obama's former pastor Jeremiah Wright. Carville went on to note, "I doubt if Governor Richardson and I will be terribly close in the future, but I've had my say... I got one in the wheelhouse, and I tagged it."

Even as Clinton's campaign began to lose steam, Carville remained both loyal and positive in his public positions, rarely veering off message and stoutly defending the candidate - but on May 13, 2008, a few hours before the primary in West Virginia, Carville remarked to an audience at Furman University in South Carolina, "I'm for Senator Clinton, but I think the great likelihood is that Obama will be the nominee." The moment marked a shift from his previous and often determinedly optimistic comments about the state of Hillary's campaign.

In 1996 Carville was inducted into the Louisiana Political Museum and Hall of Fame in Winnfield, along with former Louisiana State Treasurer Mary Evelyn Parker and the late segregationist leader Leander Perez.

Carville is married to Republican political pundit Mary Matalin, who had worked for President George H. W. Bush on his 1992 reelection campaign. Carville and Matalin were married in New Orleans in October of 1993 and now have two daughters, Matalin Mary "Matty" Carville and Emerson Normand "Emma" Carville.

In 2008 Carville and Matalin relocated their family from Virginia to New Orleans, and he is currently on the faculty of the department of political science at Tulane University.

FRANCIS H. CASE
Congressman from South Dakota

Francis Higbee Case (Dec 9, 1896 - June 23, 1962) was a journalist and politician who served for twenty-five years as a member of the United States Congress from South Dakota. He was a Republican.

Case was born in Everly, Iowa and moved with his parents to Sturgis, South Dakota at the age of thirteen. After graduating from the public schools there he attended Dakota Wesleyan University and Northwestern University, graduating in 1920. During World War I he served in the United States Marine Corps, and subsequently the United States Army Reserve and Marine Corps Reserve.

Immediately after finishing college Case began a fifteen-year career as a newspaper editor, and until 1922 he was the assistant editor of the *Epworth Herald* in Chicago, and from 1922 to 1925 he was the telegraph and editorial writer for the *Daily Journal* in Rapid City, South Dakota. From 1925 to 1931 he was editor and publisher of the *Hot Springs Star* in Hot Springs, South Dakota, and finally from 1931 until he entered Congress, he was editor and publisher of the *Custer Chronicle* in Custer, South Dakota.

Case entered politics in 1934 when he ran for a seat in the United States House of Representatives, although he lost. In 1936, however, he was elected to the U.S. House and served in it for seven terms. During that time, and before the United

States entered World War II, he was a moderate supporter of isolationism.

Case left the House in 1951 when he became a Senator. He decided to run for the Senate in 1950, and defeated incumbent John Chandler Gurney in the Republican primary. He then easily defeated Democrat John A. Engel in the general election, receiving 63% of the vote. In his first term in the Senate Case served as chairman of the United States Senate Committee on the District of Columbia from 1953 to 1955, and was a supporter of greater self-rule in the district. In 1954 he also served on a committee to investigate censuring Senator Joseph McCarthy.

Case was reelected to the Senate in 1956 in a very close race against Democrat Kenneth Holum where he received 50.8% of the vote, and he continued to serve in the Senate until his death. He died of an illness at the Naval Hospital in Bethesda, Maryland on June 23, 1962, several months before the expiration of his second term in the Senate.

Francis Case was known as a moderate Senator whose main goals were to expand America's road and waterway infrastructure, particularly in South Dakota, and as a result Lake Francis Case along the Missouri River is named after him, as is a bridge on I-395 in Washington, D.C.

JOHN CHAFEE

Governor of and Senator from Rhode Island
Secretary of the Navy

John Lester Hubbard Chafee (Oct 22, 1922 - Oct 24, 1999) was a politician who served his country as an officer in the United States Marine Corps, Governor of Rhode Island, Secretary of the Navy, and as a United States Senator.

Chafee was born in Providence, Rhode Island to a politically active family. His great-grandfather Henry Lippitt was Governor of Rhode Island (1875–1877) and among his great-uncles were Rhode Island Governor Charles Warren Lippitt and United States Senator Henry Frederick Lippitt. His uncle, Zechariah Chafee, was a Harvard law professor and a notable civil libertarian.

Chafee graduated from a coeducational primary school, Providence's Gordon School, in 1931, attended Providence Country Day School, and in 1940 graduated from Deerfield Academy in Massachusetts. He was in his third year at Yale University when the Japanese attacked Pearl Harbor, so he interrupted his undergraduate studies, enlisted in the Marine Corps, and spent his twentieth birthday fighting on Guadalcanal from August 8, 1942 until February of 1943. Then, after receiving a commission, he fought in the Battle of Okinawa as a Second Lieutenant during the spring of 1945.

Following the war Chafee received degrees from Yale University in 1947 and Harvard Law School in 1950. While

71

at Yale, he was a member of the Delta Kappa Epsilon (Phi chapter) and Skull and Bones fraternities.

In 1951 Chafee was recalled to active duty during the Korean War and served as a rifle company commander with Dog Company, 2nd Battalion, 7th Marines. Author James Brady wrote in his memoir of the Korean War of his time serving as a Marine under Chafee: "Nowhere, at any time, did John Chafee serve more nobly than he did as a Marine officer commanding a rifle company in the mountains of North Korea... he was the only truly great man I've yet met in my life..."

After Korea, Chafee became active in behind-the-scenes Rhode Island politics by helping elect a Mayor of Providence in the early 1950s. He then successfully ran for a seat in the Rhode Island House of Representatives in 1956, and later became minority leader. He was re-elected in 1958 and 1960, the latter being a year when many Republicans were swept from office in his state.

Chafee was elected Governor in 1962 and helped create the state's public transportation administration, as well as a conservation effort known as the Green Acres program. He served as Governor until 1969, when he was surprisingly defeated by underdog Democrat Frank Licht. Reasons ascribed for the defeat include the fact that he stopped campaigning after his fourteen-year-old daughter Tribbie was killed in a riding accident.

Chafee was appointed Secretary of the Navy in 1969 by President Richard Nixon. His tenure as Secretary was marked by a willingness to make bold decisions, and stand by them. Emblematic of this was his decision to elevate Admiral Elmo Zumwalt to Chief of Naval Operations over thirty-three more senior officers, and his judicious handling of the *USS Pueblo* situation. His tenure as Secretary of the

Navy is most clearly remembered for his disapproval of the recommendation to court-martial Commander Lloyd Bucher, the commanding officer of *Pueblo*. While it was clear the guilt clearly rested on the North Koreans and not Bucher or the sailors of *Pueblo*, Chafee stated that "Bucher and his men have suffered enough," and ruled that a court-martial would only add insult to injury. He served as Secretary of the Navy until 1972, when he resigned to run for the U.S. Senate.

After an unsuccessful candidacy for the Senate in 1972 against Democrat incumbent Claiborne Pell, Chafee was elected to that body in 1976 - the first Republican to win a Rhode Island Senate election since 1930. He joined the Senate Committee on Environment and Public Works in 1977 and made environmental matters a chief concern, often breaking with his party to the delight of conservation groups. He chaired that committee during his last term in office from 1995 to 1999.

Frequently following a moderate path, Chafee was pro-choice on abortion and supported the North American Free Trade Agreement. He took a moderate stance on taxes and government assistance to the needy, and on social issues Chafee was among the most liberal members of the Senate. He opposed the death penalty, school prayer, and the ban on homosexuals serving in the military. Chafee was also one of the few Republicans to support strict gun control laws. He even sponsored a bill that, if passed, would have prohibited the "manufacture, importation, exportation, sale, purchase, transfer, receipt, possession, or transportation of handguns and handgun ammunition."

During the late 1980s and 1990s Senator Chafee became an advocate for improving the U.S. health care system. He supported legislation to expand Medicaid coverage for low-income children and pregnant women, sponsored legislation to expand the availability of home and community-based services for persons with disabilities, and worked to enact legislation to establish Federally Qualified Health Centers. In 1992 he was appointed Chairman of the Senate Republican Task Force on Health, and worked to develop a consensus among Republicans on health care. In 1993 he joined with Democratic Louisiana Senator John Breaux to form the Senate Mainstream Coalition, a coalition of six Democratic and six Republican Senators seeking bipartisan consensus on health reform.

"John Chafee proved that politics can be an honorable profession," President Bill Clinton said in a statement to the Associated Press shortly after Chafee died. "He embodied the decent center which has carried America from triumph to triumph for over two hundred years."

John Chafee died suddenly from congestive heart failure in October of 1999 at the National Naval Medical Center in Bethesda, Maryland a few months after declaring he would not seek reelection in 2000. His son, Lincoln Chafee, was appointed to serve the remainder of his term. Senator Chafee was survived by his wife Virginia Coates Chafee, a daughter, and three sons.

In 2000 Senator Chafee was posthumously awarded the Presidential Medal of Freedom. In addition, the Arleigh Burke-class guided missile destroyer *USS Chafee* (DDG-90) and the John H. Chafee Blackstone River Valley National Heritage Corridor were named in his honor, and Bryant University in Smithfield, Rhode Island re-named its World Trade Center in recognition of his continuing support for global trade and the University.

MIKE COFFMAN
Congressman from Colorado

Michael "Mike" Coffman is a politician from Colorado. He is a member of the Republican Party, the current Congressman for the 6th Congressional District, and is the former Secretary of State of Colorado.

Coffman was born on March 19, 1955 in Fort Leonard Wood, Missouri to Harold and Dorothy Coffman, and is one of five children. His father served in the United States Army at Fort Leonard Wood and after 1964 at Fitzsimons Army Medical Center in Aurora, Colorado.

In 1972 Coffman enlisted in the U.S. Army and was assigned to a mechanized infantry battalion. The following year he earned a high school diploma through an Army program, and he then left active duty for the U.S. Army Reserve in 1974. Coffman then entered the University of Colorado, graduating in 1979, and also studied at Vaishnav College in Chennai, India as well as the University of Veracruz in Mexico. Upon graduation from the University of Colorado Coffman transferred from the Army Reserve to active duty in the Marine Corps and became an infantry officer. Then in 1983 he transferred from active duty to the Marine Reserves, and served until 1994.

Coffman began his political career serving as a member of the Colorado House of Representatives from 1989 to 1995. Shortly after winning reelection in 1990, he took an unpaid leave-of-absence from the State House during his active duty service in the Persian Gulf War, during which he saw combat as a light armored infantry officer. In 1994 he retired from the Marine Corps after twenty years of combined service to the Army, Army Reserve, Marines, and Marine Reserve. That same year he was elected to the Colorado State Senate, where he served as Chairman of the Finance Committee.

In 1998 Coffman was elected as State Treasurer of Colorado. He resigned from that post in 2005 in order to resume his career in the Marines and serve in the War in Iraq, where he helped support the Independent Electoral Commission of Iraq which oversaw two national elections and helped establish interim local governments in the Western Euphrates Valley. Following completion of his tour-of-duty in Iraq in 2006 he was reappointed as State Treasurer, and was subsequently elected to his next post as Colorado Secretary of State.

Coffman then announced that he would run for the U.S. House seat being vacated by Republican Tom Tancredo in 2008 in Colorado's 6th Congressional District. He won the Republican primary, and went on to defeat Democrat Hank Eng in the general election. Governor Bill Ritter then designated State Representative Bernie Buescher, a Democrat, to succeed Coffman as Secretary of State.

JON CORZINE
Governor of and Senator from New Jersey

Jon Stevens Corzine served as the fifty-fourth Governor of New Jersey from 2006 to 2010. A Democrat, Corzine had served five years of a six-year Senate term before being elected Governor in 2005. He was defeated for re-election in 2009 by Republican Chris Christie.

Corzine was born in central Illinois to Nancy June (née Hedrick) and Roy Allen Corzine on January 1, 1947. He grew up on a small family farm in Willey Station, Illinois, and after completing high school at Taylorville High School, where he had been the football quarterback and basketball captain, attended the University of Illinois at Urbana-Champaign. While in college Corzine enlisted in the Marine Corps Reserve and served from 1969 until 1975, attaining the rank of sergeant. In 1970 he enrolled in the University of Chicago Booth School of Business, from which he received a Master of Business Administration degree in 1973.

His first experience in business was in University of Chicago Booth School of Business at night. He then moved to BancOhio National Bank, a regional bank in Columbus, Ohio that was acquired by National City Bank. He worked there until 1975 when he moved his family to New Jersey and was hired as a bond trader for Goldman Sachs. Over the years he worked his way up to Chairman and CEO of the company, and in 1994 successfully converted the investment firm from a private partnership to a publicly traded corporation. Corzine's predecessor had led Goldman to its first money-losing year in its 129-year history, and to its near collapse as a firm. Corzine also chaired a presidential commission for Bill Clinton and served on the U.S. Treasury Department's borrowing committee. As a Goldman Sachs senior partner, he was summoned to help develop a rescue package for the hedge fund Long Term Capital Management when the leveraged fund's collapse in the fall of 1998 threatened the U.S. financial system. According to *U.S. News & World Report* Corzine did not get along with co-CEO Henry Paulson, and when Corzine decided to help with the bailout Paulson seized control of the firm. When Goldman Sachs subsequently went public after Corzine's departure, he made an estimated four hundred million dollars.

After being forced from Goldman Sachs in January of 1999 Corzine campaigned for a New Jersey Senate seat after Frank Lautenberg announced his retirement. Although trailing by thirty percentage points at one juncture, in the end he was elected to the Senate by a four percent margin over a four-term Republican Congressman. Corzine spent over sixty-two million of his own money on his campaign, the most expensive Senate campaign in U.S. history, with over thirty-three million spent on the primary election alone, where he defeated former Governor James Florio.

Corzine campaigned for state government programs including universal health care, universal gun registration, mandatory public preschool, and more taxpayer funding for college education. He also pushed affirmative action and same-sex marriage. David Brooks considered Corzine so liberal that although his predecessor was also a Democrat, his election helped shift the Senate to the left. During his five year senatorial career he co-sponsored 1014 bills, sponsored

145 bills (eleven of which made it out of committee), and had one sponsored bill enacted.

While in the Senate he chaired the Democratic Senatorial Campaign Committee from 2003 to 2005, and in this role was influential in convincing certain potential candidates to not run in order to avoid costly primaries in three key states during the 2004 United States Senate elections. He also played a role in the selection of Senator John Edwards as running mate for Senator John Kerry. Oddly, his resolution to congratulate Bruce Springsteen on the 30th Anniversary of *Born to Run* for his contribution to American culture was derailed, in all likelihood due to Springsteen's support of Kerry.

When Corzine ran for Governor in 2005 he continued to serve in the U.S. Senate, which ensured he could resign from the Senate and appoint a successor if he won, and retain his Senate seat if he lost. He and his opponent, Republican Doug Forrester, spent a combined seventy-three million dollars on their gubernatorial campaigns, including thirty-eight million by Corzine and nineteen million by Forrester for the general election, with the primaries accounting for the balance. Since Corzine had spent over sixty-two million on his 2000 United States Senate election, the combined expenditures for his runs for the Senate and Governorship exceeded one hundred million.

Corzine ran for re-election in 2009, and early on *Rasmussen Reports* indicated Republican challenger Chris Christie led 47% to 38%, but later polls showed Corzine closing the gap, and in some cases ahead. In the end Corzine lost the race to Christie by a margin of 48.5% to 44.9%, with 5.8% of the vote going to independent candidate Chris Daggett.

Corzine has championed expanding government health and education programs, with his plans to requiring every resident to enroll in a health plan and have taxpayers help pick up the tab for low and middle income residents. In June of 2008 state legislators voted for the first phase of that program, which mandates heath care coverage for children, and Corzine signed it into law in July.

Corzine, a death penalty opponent, presided over the abolition of capital punishment in New Jersey and replaced it with life imprisonment, making New Jersey the first state to legislatively eliminate capital punishment since 1965. Although the bill was not passed until late in 2007, New Jersey had not executed any criminals since 1963. Before the enactment of the new law, Corzine commuted the death sentences of all death row inmates to life in prison.

Jon Corzine married his kindergarten and high school sweetheart, Joanne Dougherty, in 1969 at the age of twenty-two, and their thirty-three-year marriage produced three children. The couple separated in 2002, and were divorced in November of 2003. In November 2005, Dougherty told *The New York Times* that Corzine had "let his family down, and he'll probably let New Jersey down, too." This quote was used by gubernatorial opponent Doug Forrester in a campaign advertisement.

In the spring of 1999, when Corzine was running for the United State Senate, he met Carla Katz, who was then president of Local 1034 of the Communications Workers of America, which represents the largest number of state workers in New Jersey. As Katz later recalled, Corzine offered her a job on his Senate campaign but she declined the offer. Corzine and Katz were soon dating, and they began appearing in public as a couple in early 2002, shortly after Corzine's separation from his wife. For more than two years

Corzine was romantically involved with Katz, and she lived with him at his apartment in Hoboken from April 2002 until August 2004. After Corzine's breakup with Katz their lawyers negotiated a financial payout in November of 2004. According to press accounts, the settlement for Katz exceeded six million. Corzine later admitted that he had also given $15,000 to Carla Katz's brother-in-law, Rocco Riccio, a former state employee who had resigned after being accused of examining income tax returns for political purposes. In the summer of 2005, when Corzine was running in the New Jersey gubernatorial election, news first emerged of his relationship with Katz and the money she had received. Corzine was elected Governor despite the scandal, and in the fall of 2006, during an impasse in contract negotiations between the Corzine administration and the state's seven major state employee unions (including the CWA), Katz contacted the Governor to lobby for a renewal of the negotiations. A state ethics panel, acting on a complaint from Bogota Mayor Steve Lonegan, ruled in May of 2007 that Katz's contact with Corzine during negotiations did not violate the Governor's code of conduct, but at the same time New Jersey Republican State Committee Chairman Tom Wilson filed a lawsuit requesting the release of all e-mail correspondence between Corzine and Katz during the contract negotiations. In May of 2008 New Jersey Superior Court Judge Paul Innes ruled that at least 745 pages of e-mail records should be made public, but Corzine's lawyers immediately appealed the decision.

On April 12, 2007 Governor Corzine and aide Samantha Gordon were injured in an automobile accident on the Garden State Parkway near Galloway Township while traveling from the New Jersey Conference of Mayors in

Atlantic City to his residence in Princeton to meet with radio personality (and former Marine) Don Imus and the Rutgers University women's basketball team. The New Jersey State Police determined that Corzine's SUV, driven by a state trooper, was traveling in excess of 90 MPH in a 65 MPH zone with its emergency lights flashing when the collision occurred. A pickup truck had drifted onto the shoulder and swerved back into the lane, and when another pickup truck swerved to avoid it the Governor's SUV was hit and crashed into the guardrail

Corzine and the trooper were flown by helicopter to Cooper University Hospital in Camden, a Level I trauma center, and the aide was taken by ambulance to Atlantic City Medical Center. Neither the trooper nor the aide were seriously injured, but Corzine suffered broken bones, including an open fracture of the left femur, eleven broken ribs, a broken sternum, a broken collarbone, a fractured lower vertebra, and a facial cut that required plastic surgery. The Governor was not wearing a seat belt, and friends had long said they had rarely seen him wear one. When asked why the state trooper who was driving would not have asked Corzine to put on his seat belt, a staffer said the Governor was "not always amenable to suggestion."

Corzine recuperated at his home, which had been outfitted with a videoconferencing center (at his expense) so he could communicate with legislators. He issued an apology, paid a forty-six dollar ticket (issued at the behest of his staff) for not wearing a seatbelt, and appeared in a public service announcement advocating seat belts which opened with the words "I'm New Jersey Governor Jon Corzine, and I should be dead."

JAMES DEVEREUX
Congressman from Maryland

James Patrick Sinnott Devereux (Feb 20, 1903 - Aug 5, 1988) was a Marine Corps general who was best known as the Commanding Officer of the 1st Defense Battalion during the defense of Wake Island in December of 1941. He was captured on Wake and became a prisoner of war, along with his men, after a fifteen-day battle with the Japanese. After his release in September of 1945 he concluded his military career in 1948 and went on to represent the second Congressional District of the State of Maryland in the United States House of Representatives for four terms from 1951-1959. He was also an unsuccessful candidate for Governor of Maryland in 1958.

Devereux, one of ten children, was born in Cabana, Cuba where his father was stationed as an Army surgeon. In 1910 the family moved to Chevy Chase, Maryland where Devereaux, at the age of ten, obtained a driver's license from the District of Columbia, which had no age requirement at the time.

Over the years Devereux attended the Army and Navy Preparatory School in Washington, D.C., the Tome School at Port Deposit, Maryland, LaVilla in Lausanne, Switzerland (when his parents lived in Vienna, Austria), and Loyola College of Baltimore, Maryland.

Devereux enlisted in the Marine Corps in July of 1923 at the age of twenty, was commissioned a Second Lieutenant in

February of 1925, and then was assigned to duty in Norfolk Virginia, Philadelphia Pennsylvania, Marine Barracks Quantico Virginia, and Guantanamo Bay Cuba. In 1926 he was detailed to the mail guard detachment in New York, and later was transferred to the force of Marines in Nicaragua as a company officer.

After returning to the United States early in 1927 he was assigned to serve aboard *USS Utah,* and was subsequently transferred ashore again in Nicaragua. Shortly thereafter he was ordered to the Orient and while in China was promoted to First Lieutenant, where his duties included command of the Mounted Detachment of the Legation Guard at Peking.

In 1933, following a year's tour of duty at Quantico, Devereaux was assigned to the Coast Artillery School at Fort Monroe, Virginia. Following his promotion to captain in December of 1935 he was ordered back to Quantico where he instructed at the Base Defense Weapons School and aided in the preparation of a Marine Corps manual on Base Defense Weapons.

In 1938, following a tour with the Marine Detachment aboard *USS Utah*, Devereux was transferred to the Marine Corps Base at San Diego, California. In January of 1941 he was subsequently ordered to Pearl Harbor, and later assumed command of the First Marine Defense Battalion on Wake Island. It was there, on the morning of December 8, 1941, that he received the message Pearl Harbor had been attacked by the Japanese. In the fight that followed, then-Major Devereux and his men damaged two cruisers, sank two destroyers and one escort vessel, and destroyed or damaged a total of seventy-two aircraft - and probably sank one submarine. Two more destroyers were damaged on the last day, and after fifteen days of bitter fighting 449 Marines finally surrendered to the Japanese on December 23, 1941.

After his capture Devereaux remained on Wake Island until January 12, 1942, when he was embarked along with his men on the *Nita Maru*. They stopped at Yokohama, where some American officers debarked, and later arrived at Woosung, China, which is located downriver from Shanghai, on January 24. He remained there until December of 1942, when he was transferred to Kiangwan, and was imprisoned at that camp for the next twenty-nine months before being transferred to central Hokkaidō. Devereux was finally released from the Hokkaidō Island prison camp on September 15, 1945.

After a brief rehabilitation leave Devereux was assigned as a student in the Senior Course at the Amphibious Warfare School at Quantico from September 1946 to May 1947. Upon completion of his studies he was detached to the First Marine Division at Camp Pendleton in California, and was serving there when he concluded his twenty-five-year career on August 1, 1948. In 1947 his book, *Story of Wake Island*, was published.

Devereux was advanced to the rank of brigadier general upon retirement in accordance with law, having been specially commended for his performance of duty in actual combat. For his leadership in defending the tiny American outpost for fifteen days against overwhelming odds, Devereux was awarded the Navy Cross. His citation reads in part, "For distinguished and heroic conduct in the line of his profession in the defense of Wake Island…"

Devereux took up horse farming at a farm near Glydon, Maryland, and following his retirement from the Marine Corps moved to a 200-acre farm at Stevenson, Maryland.

He was elected as a Republican to the U.S. Congress, where he served from January 3, 1951 to January 3, 1959. During his Congressional career he supported public school desegregation and ending racial discrimination in employment, and also served on the House Armed Services Committee from July 3, 1952 until he left Congress. In 1960, he was named Republican Party chairman in his district.

While stationed in the Philippines Devereux had met Mary Brush Welch, the daughter of an American missionary, and they were married in 1932. They had one son, and a daughter who died at birth in 1934. Mrs. Devereux herself died of complications from diabetes in 1942, shortly after his capture by the Japanese on Wake Island, and was buried in Arlington National Cemetery. Then in 1946 Devereaux married Rachel Clarke Cooke, and they had two sons before the second Mrs. Devereux died in 1977. He eventually married a third time, to Edna Burnside Howard, and gained a stepson and three stepdaughters.

Brigadier General James Devereux died at Stella Maris Hospice in Baltimore, Maryland on August 5, 1988 from pneumonia at the age of eighty-five. He is interred in Arlington National Cemetery.

DAVID DINKINS
Mayor of New York City

David Norman Dinkins is a former politician who was the Mayor of New York City from 1990 through 1993, as well as the first African American to hold that office.

Dinkins was born in Trenton, New Jersey on July 10, 1927 and was raised by his mother and grandmother, since his parents had divorced when he was seven years old. He attended Trenton Central High School, graduated in 1945 in the top ten percent of his class, and after graduation attempted to enlist in the United States Marine Corps. Dinkins was told that a racial quota had been filled, so he served briefly in the United States Army before joining the Marines.

Dinkins graduated magna cum laude from Howard University with a degree in Mathematics, and is a member of the Alpha Phi Alpha Fraternity, the nation's first inter-collegiate fraternity for African American men. He later graduated from Brooklyn Law School.

Dinkins rose through the Democratic Party organization in Harlem and became part of an influential group of African-American politicians which included Percy Sutton, Basil Paterson, Denny Farrell, and Charles Rangel. As an investor, Dinkins was one of fifty African-American investors who helped Percy Sutton found Inner City Broadcasting Corporation in 1971. He also served briefly in the New York State Legislature, and for many years as New York City Clerk.

Dinkins was elected Manhattan Borough President in 1985 on his third run for that office, and was elected the city's Mayor on November 7, 1989 after defeating three-term incumbent Mayor Ed Koch and two others to win the Democratic nomination and going on to narrowly defeat Rudy Giuliani, who was the Republican candidate.

Dinkins was elected in the wake of a corruption scandal that involved several Democratic leaders in New York City, and the indictment of a few key Democrats allowed him to avoid primary challenges from some potential rivals. Additionally, the fact that Dinkins is African-American helped him avoid criticism that he was ignoring the black vote by campaigning to whites.

Dinkins entered office pledging racial healing, and famously referred to New York's demographic diversity as a "gorgeous mosaic." Many New Yorkers felt that his low-key personality, which contrasted so sharply with that of his predecessor, along with the symbolism of his being the city's first black mayor, might ease racial tensions. Instead, Dinkins' term was marked by polarizing events including the 1991 Crown Heights Riot and the boycott of a Korean-owned grocery in Flatbush.

His critics have described him as weak and indecisive, but well-intentioned. Dinkins became mayor with a $1.8 billion budget deficit when he entered office, and when he left office it was $2 billion. He attempted to balance the budget by raising taxes, but high oil prices due to the Gulf War and an overall downturn in the economy did not help the economic health of the city. 300,000 private sector jobs were lost during Dinkins' administration, further eroding the city's tax base. His handling of the city's finances was also criticized as being too beholden to unions and other lobbying groups.

Dinkins' integrity came under fire, as well as his efficacy. In response to his failure to file (or pay) income taxes for five years earlier in his career, *Salon* magazine later reported that Dinkins said, "I haven't committed a crime. What I did was fail to comply with the law."

In 1991 New York was unable to pay city employees, so the Dinkins administration proposed unprecedented cuts in public services, one billion in tax increases, and the elimination of 27,000 jobs. He cut education by $579 million and marked ten homeless shelters for closing (which was opposed by the city council), but just a year later the city had a $200 million surplus.

In 1993 Dinkins lost his re-election bid to Republican Rudy Giuliani. His departure from office at the end of 1993 made him the last Democratic Mayor of New York City as of 2009, in a city where party affiliations are overwhelmingly Democratic. One factor in the loss was his perceived indifference to the plight of the Jewish community during the Crown Heights riot, and in the 1993 campaign Dinkins' support from Jews, whites, Asian Americans, and Hispanics declined substantially.

Dinkins was subsequently given a professorship at Columbia University, and although he has not attempted a political comeback he has remained somewhat active in politics. His endorsement of various candidates, including Mark J. Green in the 2001 Mayoral race, was well-publicized, and as a result he has been in conflict with Al Sharpton.

In the 2009 mayoral campaign, with Mayor Michael Bloomberg running as Independent/Republican for a third term against Democratic candidate William C. Thompson, Jr., former Mayor Giuliani campaigned for Bloomberg. On October 17 at the Jewish Community Council breakfast,

Giuliani was quoted as saying, "This city could very easily be taken back in a very different direction - it could very easily be taken back to the way it was with the wrong political leadership." The comment and others in the speech were taken as an unfavorable allusion to the Dinkins administration, although a look back suggests the comparison may not be as negative as Giuliani seemed to be implying since Dinkins did hire fellow Marine Raymond W. Kelly as police commissioner, cleanup and revitalize Times Square, and make a major commitment to rehabilitating dilapidated housing in northern Harlem, the South Bronx and Brooklyn despite significant budget constraints

David Dinkins is married to the former Joyce Burrows and they have two children. His radio program, *Dialogue with Dinkins*, can be heard Saturday mornings on WLIB radio in New York City. Dinkins also sits on the Board of Directors of The Jazz Foundation of America, and has worked to save the homes and lives of America's elderly jazz and blues musicians including those who survived Hurricane Katrina.

PAUL DOUGLAS
Senator from Illinois

Paul Howard Douglas (March 26, 1892 - September 24, 1976) was a politician and University of Chicago economist who served as a Democratic U.S. Senator from Illinois from 1949 to 1967.

Douglas was born on March 26, 1892 in the small town of Salem, Massachusetts. When he was four his mother died of natural causes, and his father remarried. His father was an abusive husband and his stepmother, unable to obtain a divorce, left her husband and took Douglas and his older brother to Onawa, Maine where her brother and uncle had built a resort in the woods.

Douglas graduated from Bowdoin College with a Phi Beta Kappa key in 1913 and then moved on to Columbia University where he earned a Master's Degree in 1915 and a Ph.D. in economics in 1921. In 1915 he married Dorothy Wolff, a graduate of Bryn Mawr College, who also earned a Ph.D. at Columbia University.

From 1915 to 1920 the Douglas' moved six times. Paul studied at Harvard University, taught at the University of Illinois, the University of Washington and Oregon's Reed College, and served as a mediator of labor disputes for the Emergency Fleet Corporation of Pennsylvania. While working for the Emergency Fleet Corporation he read John Woolman's journals, and while teaching in Seattle he joined the Quakers Religious Society of Friends.

91

In 1919 Douglas took a job teaching economics at the University of Chicago. Although he enjoyed his job, his wife was unable to obtain a position at the university due to anti-nepotism rules. When Dorothy was hired by Smith College she persuaded her husband to move the family to Amherst, where he taught at the University of Massachusetts. Douglas soon decided the situation was untenable, and in 1930 the couple divorced, with Dorothy taking custody of their four children and Douglas returning to Chicago. The following year Douglas met and married Emily Taft Douglas, who was the daughter of sculptor Lorado Taft and a distant cousin of former President William Howard Taft. Emily was a former actress, political activist, and subsequent one-term Congresswoman at-large from Illinois from 1945 to 1947.

As the 1920s drew to a close Douglas got more involved in politics. He served as an economic advisor to Republican Governor Gifford Pinchot of Pennsylvania and Democratic Governor Franklin D. Roosevelt of New York, and along with Chicago lawyer Harold L. Ickes launched a campaign against public utility tycoon Samuel Insull's stock market manipulations. He helped draft laws regulating utilities and establishing old-age pensions and unemployment insurance, and by the early 1930s was vice chairman of the League for Independent Political Action, a member of the Farmer-Labor Party's national committee, and treasurer of the American Commonwealth Political Federation.

A registered Independent, Douglas felt the Democratic Party was too corrupt and the Republican Party was too reactionary, views he expressed in a 1932 book, *The Coming of a New Party*, in which he called for the creation of a party similar to the British Labour Party. That year, he voted for Socialist candidate Norman Thomas for President of the United States.

In 1935 Douglas made his first foray into electoral politics, campaigning for the endorsement of the local Republican Party for Mayor of Chicago. Although the party endorsed someone else, Douglas continued to work with them to get their candidate in the 5th Ward elected to the city council. A strong Socialist candidate split the reform vote however, and Democrat James Cusack, a member of the Cook County political machine, was elected.

Four years later, in 1939, Cusack came up for re-election, and Douglas joined a group of reform-minded Independents who were attempting to select a suitable challenger. The group decided Douglas was the best candidate for the position, and he was summarily drafted. During the election Mayor Ed Kelly, who was in a tough fight for re-election, attempted to shore up his reputation by lending his support to Douglas' campaign. With Kelly's help, and through his own dogged campaigning, Douglas managed a narrow victory over Cusack in a runoff election.

A reformer on a council full of machine politicians and grafters, Douglas usually found himself in the minority. His attempts to reform the public education system and lower public transportation fares were met with derision, and he typically ended up on the losing end of 49-1 votes. "I have three degrees," Douglas once said after a particularly close-fought rout. "I have been associated with intelligent and intellectual people for many years. Some of these aldermen haven't gone through the fifth grade. But they're the smartest bunch of bastards I ever saw grouped together."

In 1942 Douglas officially joined the Democratic Party and ran for the United States Senate. He had the support of a cadre of left-wing activists, but the machine supported the state's at-large Congressman Raymond S. McKeough for the nomination. On the day of the primary Douglas carried 99 of the state's 102 counties, but McKeough's strong support in Cook County allowed him to win a slim majority - although he would go on to lose in the general election to incumbent Republican Senator C. Wayland Brooks.

As alderman, Douglas had worked with *Chicago Daily News* publisher Frank Knox in fighting corruption in Chicago. Knox, who had been the Republican vice-presidential nominee in 1936, had become Secretary of the Navy and was thus responsible for both the Navy and the Marine Corps.

Shortly after losing the primary Douglas resigned from the Chicago City Council, and with the aid of Knox enlisted in the Marine Corps as a private at the age of fifty. Promoted to corporal, and then to sergeant, Douglas was kept stateside writing training manuals and giving inspirational speeches to troops. He was told he was "too old to go overseas as an enlisted man," but with the aid of Knox and Knox assistant Adlai Stevenson, Douglas was eventually commissioned as an officer and subsequently sent to the Pacific theater of operations with the 1st Marine Division.

On the second day of the Battle of Peleliu Captain Douglas finally saw action when his unit waded into the fray. He received a Bronze Star for carrying ammunition to the front lines under fire, and earned his first Purple Heart when he was hit by shrapnel while carrying flamethrower ammunition. During that six week battle, while investigating some random-fire shootings, Douglas was shot at as he uncovered a two-foot-wide cave. He then killed the Japanese soldier inside, at which point he wondered whether his enemy might have been an economics professor from the University of Tokyo.

A few months later, during the Battle of Okinawa, Douglas got his second Purple Heart. While serving as a volunteer rifleman in an infantry platoon, he was helping to carry wounded from 3rd Battalion 5th Marines along the Naha-Shuri line when a burst of machinegun fire tore through his left arm and severed the main nerve, leaving it permanently disabled. Then, after a thirteen-month stay at the National Naval Medical Center at Bethesda, Douglas was discharged as a Lieutenant Colonel with full disability pay.

While Douglas had been serving in the Marines his wife Emily had been nominated to run against isolationist Republican Congressman Stephen A. Day, who had succeeded Raymond McKeough. Although she defeated Day in the 1944 election a Republican upsurge unseated her in 1946, which was the same year Douglas left the Marines.

Deciding to enter politics once again, Douglas let it be known he wished to seek the office of Governor of Illinois in 1948. Cook County machine boss Jacob Arvey, however, had a different plan. Several scandals had broken out over the machine's activities, and Arvey decided that Douglas - a scholar and war hero with a reputation for incorruptibility - would be the perfect nominee to run against Senator Brooks. Since Brooks was hugely popular in the state and had a large campaign war chest Arvey decide there was no danger of Douglas winning, so the top of the Illinois Democratic slate for the 1948 election then became Paul Douglas for Senator and Adlai Stevenson for Governor.

At the outset of the campaign Douglas' chances looked slim, but he proved to be a tenacious campaigner. He stumped across the state in a Jeep station wagon for the Marshall Plan, civil rights, repeal of the Taft-Hartley Act, more public housing, and more social security programs. During six months of non-stop campaigning he traveled more than 40,000 miles around the state and delivered more than 1,100 speeches, and when Senator Brooks refused to debate him Douglas debated an empty chair, switching from seat to seat as he provided both his own answers and Brooks'. On Election Day Douglas won an upset victory, taking fifty-five percent of the vote and defeating the incumbent by a margin of more than 407,000 votes.

Douglas soon earned a reputation as an unconventional liberal in the Senate, concerned as much with fiscal discipline as with passing the Fair Deal. Although he was a passionate crusader for civil rights (Dr. Martin Luther King Jr. described him as "the greatest of all the Senators"), Douglas earned fame as an opponent of pork barrel spending. Early in his first term he grabbed headlines when he strode to the Senate floor with magnifying glass and atlas in hand and challenged a pork barrel project for the dredging of a river in Maine by defying anyone to find the river in the atlas. When Maine's Owen Brewster objected, and pointed out the millions of dollars in pork going to Illinois, Douglas offered to cut his state's share by forty percent. He later appeared on the cover of *Time*, and a profile of him in that issue was entitled "The Making of a Maverick."

As the 1952 presidential election approached a groundswell of support arose for a Douglas candidacy for President. The National Editorial Association ranked him the second-most-qualified man after Truman to receive the Democratic presidential nomination, and a poll of forty-six Democratic insiders revealed him to be a favorite for the nomination if Truman stepped aside.

Douglas, however, refused to be considered as a candidate for President, and instead backed the candidacy of Senator Estes Kefauver of Tennessee, a folksy, coonskin cap-wearing populist who had become famous for his televised investigations into organized crime. Douglas stumped across the country for Kefauver and stood next to him at the 1952

Democratic National Convention when Kefauver was defeated by Illinois Governor Adlai Stevenson.

During the 1966 election Douglas, then seventy-four, ran for a fourth term against Republican Charles H. Percy, a wealthy businessman. A confluence of events, including Douglas' age, unhappiness within the Democratic Party over his support for the Vietnam War and open housing laws, as well as sympathy for Percy over the recent, unsolved murder of his daughter, caused Douglas to lose the election in an upset.

After losing his seat in the Senate Douglas taught at the New School, chaired a commission on housing, and wrote books, including an autobiography, *In the Fullness of Time*.

In the early 1970s Paul Douglas suffered a stroke and withdrew from public life, and on September 24, 1976 he died at his home. He was cremated, and his ashes were scattered in Jackson Park near the University of Chicago. A memorial marker at the Marine Corps Recruit Depot at Parris Island reads:

DOUGLAS VISITORS CENTER
in Memory of
SENATOR PAUL H. DOUGLAS 1892 ~ 1976

Graduating from Parris Island in 1942 as a 50 year old Private, Mr. Douglas was an inspiration to all. He rose to the rank of Major while serving in the Pacific Theater where he was wounded at Peleliu and Okinawa, and retired as a Lieutenant Colonel. The former economics professor later served as a U.S. Senator from Illinois. By his personal courage, fortitude and leadership, the Honorable Paul H. Douglas demonstrated the personal traits characteristic of Marine leaders.

JOE FOSS
Governor of South Dakota

Joseph Jacob "Joe" Foss (April 17, 1915 - January 1, 2003) was born to a Norwegian-Scots family in South Dakota, where he learned hunting and marksmanship at a young age. Like millions of others he was inspired by Charles Lindbergh, especially after he saw Lindy at an airport near Sioux Falls, and five years later he watched a Marine squadron put on a dazzling exhibition led by Captain Clayton Jerome, the future wartime Director of Marine Corps Aviation.

In 1934 Foss began his college education in Sioux Falls, but had to drop out to help his mother run the family farm when his father died. Even so he somehow managed to scrape up sixty-five dollars for private flying lessons, and five years later entered the University of South Dakota again and supported himself by waiting on tables. In his senior year he also completed a civilian pilot training program, before graduating with a Business degree in 1940.

Upon graduation Foss enlisted in the Marine Corps Reserves as an aviation cadet, and seven months later earned his Marine wings at Pensacola and was commissioned a Second Lieutenant. He was at Pensacola when the news of Pearl Harbor broke, and since he was Officer of the Day was placed in charge of base security – thus, he prepared to defend Pensacola from Jap invaders, riding around the perimeter on a bicycle. To his distress he was then ordered to aerial photographers school and assigned to VMO-1, a photo reconnaissance squadron, but insisted he wanted fighter pilot duty even after being told "You're too ancient, Joe. You're twenty-seven years old!" After lengthy lobbying with the Aircraft Carrier Training Group he learned all about the new F4F Wildcat, logging over 150 flight hours in June and July, and when he finished training became the executive officer of VMF-121. Three weeks later Foss was on his way to the South Pacific where Americans were desperately trying to turn the tide of war, and upon arrival VMF-121 was loaded aboard the escort carrier *Copahee*.

On the morning of October 9th they were catapulted off the decks in Foss' only combat carrier mission. Upon landing at Henderson Field he was told that his fighters were now based at the 'cow pasture,' and was impressed with the make-do character of the 'Cactus Air Force.' The airfield was riddled with bomb craters and wrecked aircraft, but also featured three batteries of 90mm anti-aircraft guns and two radar stations. As exec of VMF-121 he would normally lead a flight of two four-plane divisions whenever there were enough Wildcats to go around, and was the oldest pilot in the flight - four years older than the average age of twenty-three. The flight would become known as 'Foss's Flying Circus' and rack up over sixty victories, with five of his pilots becoming aces and two dying in the fight for Guadalcanal.

On October 13, 1942 VMF-121 scored its first victories when Lieutenants Freeman and Narr each got a Japanese plane. Later that same day, Foss led a dozen Wildcats to intercept thirty-two enemy bombers and fighters. In his first combat a Zero bounced Foss, but overshot, and he was able to fire a good burst and claim one destroyed aircraft. Instantly three more Zeros set upon him, and he barely made it back to 'Fighter One' with his Wildcat dripping oil.

Chastened by the experience, he declared "You can call me 'Swivel-Neck Joe' from now on." From the first day, Joe followed the tactics of Joe Bauer by getting in close - so close that another pilot joked the 'exec' left powder burns on his targets. The next day, while intercepting a flight of enemy bombers, Foss' engine acted up and he took cover in the clouds - but suddenly a Wildcat whizzed past him tailed by a Zero. Foss cut loose and shot the Zero's wing off, and scored his second victory in two days.

While the Wildcats' primary responsibility was air defense, they also strafed Japanese infantry and ships when they had enough ammunition. Joe led one such mission on the 16th. Mid-October was the low point for the Americans in the struggle for Guadalcanal. Japanese warships shelled U.S. positions nightly, with special attention to the airstrip. To avoid the shelling, some fliers slept on the front lines.

Foss was leading an interception on morning of the 18th when the Zero top cover pounced on them and downed an F4F, but he was able to get above them and flame the nearest, hit another, and briefly engage a third. Gaining an angle, he finally shot up the third plane's engine. Next he found a group of Bettys already under attack by VF-71. He executed a firing pass from above, flashed through the enemy bombers, and pulled up sharply, blasting one from below. Nine days at Guadalcanal, and he was an ace! Two days later Lieutenant Colonel Bauer and Foss led a flight of Wildcats on the morning intercept. In the dogfighting Foss downed two Zeros, but took a hit in his engine and landed safely at Henderson Field with a bad cut on his head.

On the 23rd 'Cactus' put up two flights, led by Foss and Major Davis. There were plenty of targets, and Foss soon exploded a Zero. He went after another which tried to twist away in a looping maneuver, so he followed and opened up while inverted at the top of his loop. He caught the Zero and flamed it, later describing it as a lucky shot. Next Foss spotted a Japanese pilot doing a slow roll. He fired as the Zero's wings rolled through the vertical and saw the enemy pilot blown out of the cockpit - minus a parachute. Suddenly Foss was all alone and two Zeros hit him, but his rugged Grumman absorbed the damage, permitting him to flame one of his assailants. Once again, he nursed a damaged fighter back to Guadalcanal. So far he had destroyed eleven enemy planes, but had brought back four Wildcats which were too damaged to fly again.

October 25th was the day the Japanese planned to occupy Henderson Field, and they sent their fighters over with orders to circle until the airstrip was theirs. It didn't work out that way, as U.S. ground forces held their lines and 'Cactus did its part. Joe Foss led six Wildcats up before 10 AM, and claimed two of the Marines' three kills on that sortie. In an afternoon mission on the 25th he downed three more, to become the Marine Corps' first 'ace in a day.' Foss had achieved fourteen victories in only thirteen days.

On November 7th Foss led seven F4Fs up the 'Slot' to attack some destroyers and a cruiser which were covered by six Rufe floatplane fighters. They dispatched five of the Rufes promptly, and prepared to strafe the destroyers as Foss climbed up to protect the others and got involved in a dogfight with a Pete, a two-man float biplane. He shot down the slow-flying plane, but not before its rear gunner perforated the Wildcat's engine with machine gun fire. Once again Foss' aircraft sputtered on the way home, but his time it didn't make it. As the engine died he put it into the longest possible shallow dive, to get as close to home as he could.

As the plane went into the water off Malaita Island Foss struggled with his parachute harness and seat. He went under with his plane, gulped salt water, and almost drowned before he freed himself and inflated his Mae West. Exhausted and with the tide against him, he knew that he couldn't swim to shore. While trying to rest and re-gain his strength in his life raft, he spotted shark fins nearby. He sprinkled the chlorine powder supplied for that purpose in his emergency pack, and that seemed to help. Then, as darkness approached, he heard some searchers looking for him. They hauled him in and brought him to Malaita's Catholic mission, where there were a number of Europeans and Australians, including two nuns who had been there for forty years and had never seen an automobile.

The next day a PBY Catalina rescued him, and on his return to Guadalcanal he learned 'Cactus' had downed fifteen Japanese planes in the previous day's air battle. His own tally stood at nineteen, and on the 9th of November Admiral Bull Halsey pinned the Distinguished Flying Cross on Foss and two other pilots.

The Americans were bringing four transports full of infantry to Guadalcanal three days later, and the Japanese sent sixteen Betty bombers and thirty covering Zeroes after them while American Wildcats and Airacobras defended. Foss and his Wildcats were flying top cover CAP and dove headlong into the attackers, right down onto the deck. As Barrett Tillman described it in *Wildcat Aces of WWII:* "Ignoring the peril, Foss hauled into within one hundred yards of the nearest bomber and aimed at the starboard engine, which spouted flame. The G4M tried a water landing, caught a wingtip and tumbled into oblivion. Foss set his sight on another Betty when a Zero intervened. The F4F nosed up briefly and fired a beautifully aimed snapshot which sent the A6M spearing into the water. He then resumed the chase."

Foss caught up with the next Betty in line and made a deflection shot into its wingroot, and the bomber flamed up and set down in the water. The massive dogfight continued, until everyone finally ran out of fuel and ammunition. Late that afternoon Colonel Bauer, tired of being stuck on the ground at Fighter Command, went up with Foss to take a look. It was his last flight, and was described by Joe Foss in a letter to Bauer's family. No trace of 'Indian Joe' was ever found. Then, back at Guadalcanal, Foss was diagnosed with malaria. The two great leaders of Cactus Fighter Command were gone, although Foss would return in six weeks.

Foss returned to combat flying on the 15th when he shot down three more planes to bring his total to twenty-six. He flew his last mission ten days later when his flight and four P-38s intercepted a force of over sixty Zeros and Vals. Quickly analyzing the situation, he ordered his flight to stay high, circling in a Lufbery. This made his small flight look like a decoy to the Japanese. Soon Cactus scrambled more fighters, and the Japanese planes fled. It was ironic that in one of Joe Foss' most satisfying missions, he didn't fire a shot.

A few months later Foss went to Washington D.C. to be decorated and begin "the dancing bear act" for his twenty-six aerial victories - which had equaled Eddie Rickenbacker's World War One record. He gave pep talks, made factory tours, and went on the inevitable War Bond tours. Then in May of 1943 President Roosevelt presented him with the Medal of Honor for outstanding heroism above and beyond the call of duty.

After the war Foss was commissioned in the South Dakota Air National Guard, which he helped organize. He then turned to politics and was elected to the South Dakota House of Representatives, and during the Korean War returned to active duty as an Air Force Colonel. Foss later became the chief of staff of the South Dakota Air National Guard with the rank of Brigadier General, and in 1954 was overwhelmingly elected Governor of South Dakota. Two years later he was elected to a second term, and once that expired Foss was elected the first commissioner of the American Football League and served in that capacity until 1966. He was also president of the National Rifle Association (NRA) from 1988 - 1990, and was featured in Tom Brokaw's best-seller *The Greatest Generation.* Joe Foss was *much* more than "just" a fighter pilot!

On January 11, 2002 Foss, then in his mid-80s, gained renewed fame when he was stopped at the Phoenix Sky Harbor International Airport because he was carrying his Medal of Honor (which has pointed edges), along with a small pocketknife (with MOH insignia) on his way to giving a speech to a class at the United States Military Academy at West Point. The subsequent delay and lack of recognition of the award, together with his age, were used as an example of alleged widespread abuse of passengers by airport security personnel pre-TSA.

Foss coauthored or was the subject of three books, including the wartime *Joe Foss: Flying Marine* (with Walter Simmons), *Top Guns* (with Matthew Brennan), and *A Proud American* by his wife, Donna Wild Foss. Foss also provided the foreword to *Above and Beyond: the Aviation Medals of Honor* by Barrett Tillman.

Joe Foss died on New Year's Day 2003 after suffering a severe stroke three months previously. He was buried at Arlington National Cemetery on January 21, 2003, and his name and patriotic activities are perpetuated at the Foss Institute in Scottsdale, Arizona.

ORVILLE FREEMAN
Governor of Minnesota

Orville Lothrop Freeman (May 9, 1918 – Feb 20, 2003) was a politician who served as the 29th Governor of Minnesota from January 5, 1955 to January 2, 1961, and as U.S. Secretary of Agriculture from 1961 to 1969 under Presidents John F. Kennedy and Lyndon B. Johnson. He was one of the founding members of the Minnesota Democratic-Farmer-Labor Party, and was strongly influential in the merger of the pre-DFL Minnesota Democratic and Farmer-Labor Parties.

Born in 1918 in Minneapolis, Minnesota of Swedish and Norwegian ancestry, Freeman is best remembered for initiating the Food Stamp Program for under-resourced people which is still in use today. He was a 1940 graduate of the University of Minnesota, where he met life-long friend and political ally Hubert Humphrey. During World War II Freeman served as a combat officer in the United States Marine Corps, and achieved the rank of major.

Figuring that the United States was going to be getting involved in World War II, Freeman signed up for the Marine Reserves in late 1940 with the understanding he could finish law school before fulfilling his required service. The attack on Pearl Harbor changed all that, and on December 31, 1941 he received orders to report to Officer Candidate School at Marine Corps Base Quantico.

After graduating from OCS and follow-on training to be an infantry officer, he reported to Camp Elliot near San Diego, California and was soon assigned to Kilo Company, 3rd Battalion, 9th Marines. His unit would eventually ship out for a period of training in New Zealand, and was later deployed to Guadalcanal.

On November 1, 1943 Freeman saw his first combat when his unit came ashore at Torokina on Bougainville in what would turn out to be the first battle of the Bougainville Campaign. A few days later, while leading a patrol, he came across a group of five or six Japanese soldiers in a clearing. Freeman was able to shoot a few of them, but was also shot himself in the jaw and left arm. He was evacuated to an Army hospital on New Caledonia and later a Naval hospital on Noumea, and returned to the United States in 1944 - but was never able to recover enough movement in his arm to pass a Marine Corps physical and get back to combat.

Once his recovery was complete Freeman earned his LL.B. from the University of Minnesota Law School in 1946 and went on to practice law in Minneapolis. He ran unsuccessfully for Attorney General of Minnesota in 1950 and for Governor in 1952, but was elected Governor in 1954 and re-elected in 1958. He also later served in Washington as Secretary of Agriculture While serving as Governor, on November 13, 1955, Freeman was a guest on the variety show *Toast of the Town* (which would later be called *The Ed Sullivan Show*).

Orville Lothrop Freeman died from complications of Alzheimer's disease in February of 2003 in Minneapolis, Minnesota at the age of eighty-four.

WAYNE GILCHREST
Congressman from Maryland

Wayne Thomas Gilchrest is a former member of the United States House of Representatives who represented Maryland's 1st Congressional District. In 2008, the moderate Gilchrest was defeated in the Republican primary by State Senator Andy Harris.

Gilchrest was born on April 15, 1946 in Rahway, New Jersey, the fourth of Elizabeth and Arthur Gilchrest's six boys. After graduating from high school in 1964 he joined the Marine Corps, and during his tour of duty saw action during the invasion of the Dominican Republic and ultimately in the Vietnam War. He earned the rank of Sergeant in Vietnam where, as a platoon leader, he was wounded in the chest. Gilchrest was decorated with the Purple Heart, Bronze Star, and Navy Commendation Medal, and today he is a member of the American Legion, Veterans of Foreign Wars, and Military Order of the Purple Heart.

In 1969 Gilchrest received an Associate's Degree from Wesley College in Dover, Delaware, and then spent a semester in Kentucky studying rural poverty in Appalachia. He went on to receive a Bachelor's Degree in history from Delaware State College in 1973, and since then has done work towards a Master's Degree at Loyola College in Baltimore.

While teaching at Kent County High School on the Eastern Shore, Gilchrest ran against four-term 1st District Democratic incumbent Roy Dyson in 1988, who was plagued by allegations of improper contributions from defense contractors and questions about his sexual orientation. Despite being badly outspent Gilchrest nearly unseated Dyson, losing by only 460 votes. He sought a rematch in 1990, and this time soundly beat Dyson by fourteen percentage points. Then in 1992 he survived a close contest against Tom McMillen, who had represented the 4th District before being drawn into the 1st District. Gilchrest won by only three percent, but wouldn't face serious opposition again for over a decade.

Gilchrest is a member of many moderate Republican groups such as the Republican Main Street Partnership, Republicans for Environmental Protection, and the Republican Majority For Choice. He was also the co-chairman of the Congressional Climate Change Caucus together with Democrat John Oliver of Massachusetts. Gilchrest was a Republican co-sponsor of Representative Marty Meehan's "Military Readiness Enhancement Act" which would have repealed the "Don't ask, don't tell" policy. He also spoke in favor of same-sex marriage while the Maryland Legislature was considering legalizing it, calling it a matter of "social justice, civil rights and a more viable democracy."

Aside from his socially moderate stance, Gilchrest has drawn attention for his stance on the Iraq War. Though he initially supported it, his support waned as the occupation became increasingly violent, and he expressed his support for the Iraq Study Group Report and called for setting a timetable for withdrawal. Gilchrest also joined sixteen Republicans and 229 Democrats by voting in favor of House Concurrent Resolution 63, which was a non-binding resolution expressing disapproval for the Iraq War troop surge of 2007.

Gilchrest's moderate voting record resulted in vigorous primary challenges from Republicans who considered him a "Republican in Name Only," however none were successful until 2008. That year State Senator Andrew Harris, State Senator E. J. Pipkin, Joe Arminio and Robert Banks challenged Gilchrest in the Republican primary, with Harris being strongly supported by the Club for Growth. Harris defeated Gilchrest in the Republican primary, with Pipkin finishing third. After Gilchrest' loss he broke with his party and endorsed Queen Anne's County State's Attorney Frank Kratovil, the Democratic nominee, in the general election and was quoted as saying, "Let's see, the Republican Party, or my eternal soul?" and "Party loyalty, or integrity?" when questioned. On September 18, 2008 Gilchrest made radio comments praising the Democratic Presidential ticket of Barack Obama and Joe Biden, causing some media outlets to claim his endorsement of the Democratic ticket. Gilchrest quickly clarified these comments, saying that they did not amount to an endorsement. Despite the fact that he did not officially endorse Obama, in an October *Washington Post* article Gilchrest sharply criticized his own party and their presidential nominee, fellow Vietnam veteran John McCain. He said that the Republican Party "has become more narrow, more self-serving, more centered around "I want, I want, I want." and said McCain "recites memorized pieces of information in a narrow way, whereas Barack Obama is constantly evaluating information and using his judgment. One guy just recites what's in front of him, and the other has initiative and reason and prudence and wisdom."

JOHN GLENN
Senator from Ohio

John Herschel Glenn, Jr. is a retired United States Marine Corps pilot, a former astronaut and United States Senator, and the first American and third person to orbit the Earth. Glenn was a Marine Corps fighter pilot before joining NASA's Mercury program as a member of the original astronaut group, and orbited the Earth in *Friendship 7* in 1962. After retiring from NASA he entered politics as a Democrat, and represented Ohio in the United States Senate from 1974 to 1999.

Glenn received a Congressional Space Medal of Honor in 1978 and was inducted into the Astronaut Hall of Fame in 1990. In 1998 he became the oldest person to fly in space, and the only one to fly in both the Mercury and Shuttle programs, when at age seventy-seven he flew on Space Shuttle *Discovery*. Glenn and M. Scott Carpenter are the last surviving members of the Mercury Seven.

John Glenn was born on July 18, 1921 in Cambridge, Ohio to John Herschel Glenn and his wife Clara Theresa (née Sproat) Glenn and was raised in New Concord, Ohio. Glenn studied chemistry at Muskingum College, and received his private pilot's license as a physics course credit in 1941. When the attack on Pearl Harbor brought the United States into World War II he dropped out of college and enlisted in the U.S. Army Air Corps, however the Army did not call him up so in March of 1942 he enlisted as a United

States Navy aviation cadet. Glenn trained at Naval Air Station Olathe, where he made his first solo flight in a military aircraft, and in 1943 during advanced training at Naval Air Station Corpus Christi he was reassigned to the Marine Corps. It was during this period, on April 6, 1943, that Glenn married his childhood sweetheart, Anna Margaret Castor. They had met in New Concord and had played together in the school band. After completing his training the newly married Glenn was assigned to Marine squadron VMJ-353, flying R4D transport planes. He eventually managed a transfer to VMF-155 as an F4U Corsair pilot, and flew fifty-nine combat missions in the South Pacific. Glenn saw action over the Marshall Islands, where he attacked anti-aircraft batteries and dropped bombs on Maloelap. Then in 1945 he was assigned to Naval Air Station Patuxent River, Maryland, and was promoted to captain shortly before the war ended.

After the war Glenn flew patrol missions in North China with VMF-218 until his squadron was transferred to Guam. He then became a flight instructor at Naval Air Station Corpus Christi, Texas in 1948, attended the amphibious warfare school, and received a staff assignment.

Glenn was next assigned to VMF-311, flying the new F9F Panther jet interceptor. He flew his Panther on sixty-three combat missions during the Korean War, gaining the dubious nickname "Magnet Ass" from his apparent ability to attract enemy flak - twice he returned to base with over 250 flak holes in his aircraft. Glenn flew for a time with Ted Williams, the future Hall of Fame baseball player for the Boston Red Sox, as his wingman.

Glenn later flew a second Korean combat tour on an inter-service exchange program with the United States Air Force. He logged twenty-seven missions in the faster F-86 Sabre, and shot down three MiG-15s near the Yalu River in the final days before the cease fire.

Glenn returned to Naval Air Station Patuxent River after being appointed to the Test Pilot School where he served as an armament officer, flying planes to high altitude and testing their cannons and machine guns. Then on July 16, 1957 Glenn completed the first supersonic transcontinental flight in a Vought F8U-1 Crusader. The flight from NAS Los Alamitos, California to Floyd Bennett Field, New York took 3 hours, 23 minutes and 8.4 seconds, and as he passed over his hometown a child in the neighborhood reportedly ran to the Glenn house shouting "Johnny dropped a bomb! Johnny dropped a bomb! Johnny dropped a bomb!" as the sonic boom shook the town. Project Bullet, the name of the mission, included both the first transcontinental flight to average supersonic speed (despite three in-flight refuelings during which speeds dropped below 300 mph), and the first continuous transcontinental panoramic photograph of the United States. Glenn received his fifth Distinguished Flying Cross for the mission.

In April of 1959, despite the fact that Glenn had failed to earn the required college degree, he was assigned to the National Aeronautics and Space Administration (NASA) as one of the original group of astronauts for the Mercury Project. During this time he remained an officer in the Marine Corps, and became the fifth person in space and the first American to orbit the Earth. While aboard *Friendship 7* on February 20, 1962, on the "Mercury Atlas 6" mission, Glenn circled the globe three times during a flight lasting 4 hours, 55 minutes, and 23 seconds. During the mission there was concern that his heat shield had failed and his craft would burn up on re-entry, but he made his splashdown safely. Glenn was celebrated as a national hero, and received a ticker-tape parade reminiscent of Lindbergh.

In July of 1962 Glenn testified before the House Space Committee in favor of excluding women from the NASA astronaut program. The impact of such testimony from so prestigious a national hero is debatable, but no female astronaut flew on a NASA mission until Sally Ride in 1983, and none piloted a mission until Eileen Collins in 1995 - more than thirty years after the hearings.

Six weeks after the assassination of President John F. Kennedy Glenn resigned from NASA to run for office in his home state of Ohio, and in 1965 he retired from the Marine Corps as a Colonel. Some accounts of Glenn's years at NASA suggest he was prevented from flying Gemini or Apollo missions, either by President Kennedy or NASA management, on the grounds that the loss of a national hero of his stature would seriously harm or even end the manned space program - yet Glenn resigned from the astronaut corps in January of 1964, well before the first Gemini crew was assigned.

Three decades later, after serving twenty-four years in the Senate, Glenn lifted off for a second space flight on October 29, 1998 aboard Space Shuttle *Discovery* in order to study the effects of space flight on the elderly. At age seventy-seven, John Glenn became the oldest person ever to go into space. It is interesting to note that prior to Glenn the record had been held by sixty-two-year-old former Marine Story Musgrave, and before him the oldest man in space had been fifty-nine-year-old Vance Brand, who had also served in the Marine Corps.

Glenn's participation in the nine-day mission was criticized by some in the space community as a junket for a politician, while others noted that the flight offered valuable research on weightlessness and other aspects of space flight on the same person at two points in life thirty-six years apart. Upon the safe return of STS-95 Glenn (and his crewmates) received another ticker-tape parade, making him the ninth (and, as of 2007, latest) individual to have received multiple ticker-tape parades in his lifetime.

In 1970 Glenn contested for the Democratic nomination for U.S. Senate but was defeated in the primary by fellow Democrat Howard Metzenbaum, who went on to lose the general election race to Robert Taft Jr. Then, in a bitterly-fought 1974 Democratic primary rematch, Glenn defeated Metzenbaum, who had been appointed by Ohio Governor John Gilligan to fill out the Senate term of William B. Saxbe after the latter resigned to become U.S. Attorney General. In the general election Glenn defeated the Republican Mayor of Cleveland, Ralph Perk, and began a Senate career that would continue until 1999.

In 1976 Glenn was a candidate for the Democratic Vice Presidential nomination, but his keynote address at the party's National Convention failed to impress the delegates and the nomination went to veteran politician Walter Mondale. Glenn also mounted a bid to be the 1984 Democratic Presidential candidate and polled well, coming in a strong second to Mondale. It was also surmised that he would be aided by the almost-simultaneous release of *The Right Stuff,* a film about the original seven Mercury astronauts in which Glenn's character was portrayed (by actor Ed Harris) in an appealing manner, but Glenn thought it would be bad form to capitalize on this kind of publicity and didn't make much of it during the period leading up to the Iowa caucuses. Media attention turned to Mondale, Gary Hart, and Jesse Jackson, and by the time Glenn's campaign started playing up *The Right Stuff* for the New Hampshire primary it was too late.

During his time in the Senate Glenn was chief author of the 1978 Nonproliferation Act, served as chairman of the Committee on Governmental Affairs from 1987 until 1995, and sat on the Foreign Relations and Armed Services committees and the Special Committee on Aging. Once Republicans regained control of the Senate he also served as the ranking minority member on a special Senate investigative committee chaired by Tennessee Senator Fred Dalton Thompson which looked into alleged illegal donations by China to U.S. political campaigns during the 1996 election.

Glenn helped found the John Glenn Institute for Public Service and Public Policy at the Ohio State University, and in July of 2006 the institute merged with OSU's School of Public Policy and Management to become the John Glenn School of Public Affairs. Today he holds an adjunct professorship at both the Glenn School and in Ohio State's Department of Political Science.

On September 5, 2009 John and Annie Glenn "dotted the 'i'" during The Ohio State University's Script Ohio Marching Band performance at halftime during the Ohio State vs. Navy football game. Bob Hope, Woody Hayes, Buster Douglas, Dr. E. Gordon Gee, Novice Fawcett, Robert Ries and Jack Nicklaus are the only other non-band members to have received this honor.

John Glenn was inducted into the Astronaut Hall of Fame in 1990, and the NASA John H. Glenn Research Center at Lewis Field in Cleveland, Ohio is named after him. The Senator John Glenn Highway runs along a stretch of I-480 across from the Center, and the Colonel Glenn Highway runs by Wright-Patterson Air Force Base outside Dayton, Ohio.

John Glenn High School in his hometown of New Concord, Ohio and Colonel John Glenn Elementary in Seven Hills, Ohio were named for him as well. In 2004, Colonel Glenn was awarded the Woodrow Wilson Award for Public Service by the Woodrow Wilson International Center for Scholars of the Smithsonian Institution.

HOWELL HEFLIN
Senator from Alabama

Howell Thomas Heflin (June 19, 1921 - March 29, 2005) was a United States Senator from Alabama and a member of the Democratic Party.

Heflin was born in Poulan, Georgia, the nephew of prominent white supremacist politician James Thomas Heflin and great nephew of Alabama Congressman Robert Stell Heflin. He attended public school in Alabama, and earned his Bachelor of Arts degree in 1942 from Birmingham-Southern College.

During World War II, from 1942 to 1946, Heflin served as an officer in the Marine Corps and was awarded the Silver Star for valor in combat as well as two Purple Heart medals after being wounded in action on Bougainville and Guam.

After World War II Heflin attended Law School at the University of Alabama, graduating in 1948. He later became a law professor, and was then Chief Justice of the Alabama Supreme Court from 1971 to 1977.

In 1978 Heflin was elected as a Democrat to the United States Senate to succeed John Sparkman. He rose to become Chairman of the Select Committee on Ethics and held that post until January of 1997, during which time he led the prosecution against fellow Senator Howard Cannon (D-NV) for violations of Senate rules.

His stances on cultural issues most often reflected his religious views. Heflin strongly opposed legal abortion and all gun control laws, supported prayer in public schools, and opposed extending federal laws against discrimination to lesbians and gays. He voted in favor of the Gulf War and against limiting spending on defense, and along with Fritz Hollings of South Carolina was one of only two Democrats in the Senate to vote against the Family and Medical Leave Act. Heflin occasionally voted with Republicans on taxes, but on other economic issues he was more in sync with the populist wing of his party. He voted against the North American Free Trade Agreement (NAFTA), General Agreement on Tariffs and Trade (GATT), and attempts to weaken enforcement of consumer protection measures. He strongly supported affirmative action laws, and memorably voted against the nomination of Clarence Thomas for the Supreme Court.

During his tenure in the Senate Heflin was widely considered to have bipartisan support if he were nominated by President Reagan to fill a vacancy on the United States Supreme Court, but said he did not wish to serve on the highest court in the land.

In July of 1994 Senator Heflin was dining in the Capitol with some Alabama reporters, and suddenly felt like he had to sneeze. He reached into his pocket and pulled out a bit of fabric and began to wipe his nose with what turned out to be a pair of ladies underwear. His office later released the following press release: "(This morning) I mistakenly picked up a pair of my wife's white panties and put them in my pocket while rushing out the door to go to work. Rather than take a chance on being embarrassed again, I'm going to start buying colored handkerchiefs."

Heflin was also known to have a sharp sense of humor. Upon seeing photos in the *National Enquirer* showing Senator Ted Kennedy copulating with an unknown woman on the deck of Kennedy's boat, Heflin commented, "Well, I declare. I do believe the Senator from Massachusetts has changed his position on offshore drillin'!"

Senator Howell Heflin died on March 29, 2005 of a heart attack at the age of eighty-three. The University of Alabama School of Law has since honored him with the Howell Heflin Conference Room in the Bounds Law Library, and the Howell T. Heflin Seminar room in the Library of Birmingham-Southern College is named in his honor as well. In addition there is now a street named Howell Heflin Lane in Tuscumbia, Alabama, and the Howell Heflin Lock and Dam is named for him.

DUNCAN D. HUNTER
Congressman from California

Duncan Duane Hunter is a Republican member of the United States House of Representatives for California's 52nd Congressional District and the son of his predecessor, Duncan Hunter. The district is located in northern and eastern San Diego County and includes El Cajon, La Mesa and a portion of eastern San Diego.

Hunter is a veteran of both the Iraq War and the War in Afghanistan, and along with John Boccieri (D-OH) is one of only two members of Congress to have served in both conflicts. He is now the only combat veteran of either war still serving.

Duncan Hunter was born on December 7, 1976 in San Diego, California and graduated from Granite Hills High School in El Cajon. He then attended San Diego State University, where he earned a degree in Business Administration. He worked to pay for his college education by creating websites and programming databases and ecommerce systems for high-tech companies, and upon graduating from San Diego State went to work full time in San Diego as a business analyst.

The day after the September 11 terrorist attacks in 2001 Hunter quit his job and joined the Marine Corps. He attended Officer Candidates School at Marine Corps Base Quantico, and was commissioned a Second Lieutenant in March of 2002. He went on to serve as a field artillery officer in the

1st Marine Division during the 2003 invasion of Iraq, and completed a second tour in Fallujah in 2004 with Battery A, 1st Battalion, 11th Marine Regiment during which time he participated in Operation Vigilant Resolve. In September of 2005 Hunter was honorably discharged from active duty and began life as a civilian, but remained in the Marine Corps Reserve. In 2007 he was re-called to active duty for a tour of duty in Afghanistan in support of Operation Enduring Freedom, was honorably discharged from active duty in December of that year, and today continues to serve as a Captain in the Marine Corps Reserve.

Hunter won the Congressional Republican primary in June of 2008 by receiving seventy-two percent of the vote, and in November defeated Democratic opponent Mike Lumpkin by a fifty-seven to thirty-nine percent margin. In doing so he replaced his father, Congressman Duncan L. Hunter, the former Chairman of the House Armed Services Committee and a former Republican candidate for President, who had retired from Congress after fourteen terms.

Following in his father's footsteps, Hunter's voting record has been decidedly conservative. He is a member of the conservative Republican Study Committee, of which his father was also a member, and both he and his father have signed the Taxpayer Protection Pledge.

BILL JANKLOW
Governor of South Dakota

William John "Bill" Janklow is the former Republican Governor of South Dakota, and also served as a Congressman in the House of Representatives for just over a year before resigning after being convicted of vehicular manslaughter. He is currently a lawyer and lobbyist.

Janklow was born on September 13, 1939 in Chicago, Illinois and served in the Marine Corps from 1956 to 1959. He later graduated from the University of South Dakota in 1964 with a BS in business administration, and received a law degree in 1966. After graduating from law school he was a Legal Services lawyer for six years on the Rosebud Indian Reservation, and in 1973 was appointed the Chief Prosecutor of South Dakota and quickly earned a reputation as a top trial lawyer.

After serving as South Dakota's Attorney General from 1975 to 1979 Janklow was elected Governor in 1978, and was easily reelected in 1982 with over seventy percent of the vote - the highest percentage for a gubernatorial candidate in the state's history.

One of his first acts as Governor was signing into law a bill reinstating capital punishment, and another major action of his administration was dropping South Dakota's cap on interest rates. That allowed Citibank to open a credit card center in Sioux Falls from which it could charge high rates.

Several states had similar usury laws, but under federal banking rules a state had to formally invite a bank into their jurisdiction - and South Dakota was able to invite Citibank before the others.

Barred by state law from running again in 1986, Janklow challenged incumbent U.S. Senator James Abdnor in the Republican primary. Janklow lost, but the bruising primary battle weakened Abdnor and contributed to the latter's loss in the general election to Democrat Tom Daschle, who was at the time South Dakota's lone member of the U. S. House of Representatives.

Janklow returned to politics in 1994 when he defeated incumbent Governor Walter Dale Miller in the Republican primary. He was handily elected that year, and reelected in 1998. In his second two terms Janklow cut property taxes for homeowners and farmers by thirty percent, and made up the revenue loss caused by the voters repealing the inheritance tax. In the end he was the longest serving Governor in South Dakota history, and the only person in the state's history to serve eight full years as Governor - which he did twice.

In 2002 Janklow ran for South Dakota's only House seat and defeated Democrat Stephanie Herseth, an attorney and the granddaughter of former Governor Ralph Herseth and his wife, former South Dakota Secretary of State Lorna Herseth.

Then, in August of 2003, Janklow was involved in a fatal traffic collision when he struck and killed motorcyclist Randolph Scott at a rural intersection near Trent, South Dakota. Scott was thrown from his motorcycle and killed instantly, and Janklow suffered a broken hand and bleeding on the brain. In the ensuing investigation it was determined Janklow was likely driving at least seventy miles per hour in a fifty-five mph zone, and that he had run a stop sign at the intersection where the crash occurred.

Janklow was arraigned on August 29, and said he "couldn't be sorrier" for the accident. At trial his defense lawyer argued that he had suffered a bout of hypoglycemia, or low blood sugar, and was thus "confused," and Janklow testified that he had taken an insulin shot the morning of the accident and had subsequently not eaten anything throughout day.

Janklow was convicted by a Moody County jury of second-degree manslaughter, and a few days later resigned his seat in Congress because the conviction substantially limited his role until the House Ethics Committee could conduct an investigation.

From early on in his political career Bill Janklow was someone people either loved or hated. Dubbed by some as the "pirate saint," Janklow amassed a fairly impressive list of achievements on behalf of the people of South Dakota during his sixteen years as their chief executive. Although controversial, he is among the most electorally successful politicians in South Dakota's history after being elected to statewide office six times.

Bill Janklow began to work as an attorney once again in 2006 after the South Dakota Supreme Court granted a petition for the early reinstatement of his law license. Soon afterwards the Mayo Clinic retained him to lobby against the DM&E Railroad expansion, and he also represents landowners who are seeking reimbursement from the railroad for property seized under imminent domain. Approximately fifty percent of those cases are handled pro bono.

TED KULONGOSKI
Governor of Oregon

Theodore R. "Ted" Kulongoski is a politician who is currently serving his second term as the 36th Governor of Oregon. A Democrat, he has served in both houses of the Oregon Legislative Assembly, as State Insurance Commissioner, as Attorney General, and as an Associate Justice on the Oregon Supreme Court.

Kulongoski was born in rural Missouri on November 5, 1940. He was four years old when his father died, and spent the rest of his childhood in a Catholic boys' home. After high school Kulongoski served in the Marines, and with the help of the G.I. Bill he obtained both undergraduate and law degrees from the University of Missouri. He then moved to Eugene, Oregon and became a labor lawyer.

In 1974 Kulongoski was elected to the Oregon House of Representatives, and in 1978 to the Oregon State Senate. Then in 1980 he ran for the United States Senate but lost to incumbent Republican Bob Packwood, and in 1982 he made a bid for Governor but was defeated by Republican incumbent Victor Atiyeh.

In 1992 Kulongoski was elected Oregon Attorney General and focused on reforming the juvenile justice system, but in 1996 decided against running for re-election and instead successfully ran for the Oregon Supreme Court. Then five years later, he resigned from the court to run for Governor.

Kulongoski narrowly won that election, and took office on January 13, 2003. He inherited a state facing a massive budget deficit and high unemployment, and also had the task of fixing the public employees' pension system without angering the labor unions which had backed his campaign.

Kulongoski ran for reelection four years later, and defeated multiple opponents in the general election - Republican candidate Ron Saxton, Constitution Party candidate Mary Starrett, Libertarian Party candidate Richard Morley, and Pacific Green Party candidate Joe Keating

MIKE MANSFIELD
Congressman and Senator from Montana
Ambassador to Japan

Michael Joseph "Mike" Mansfield (March 16, 1903 - October 5, 2001) was a Democratic politician and the longest-serving Majority Leader in the history of the United States Senate, serving from 1961 to 1977, and also served as United States Ambassador to Japan for ten years. Although Mansfield was born in New York City to Irish Catholic immigrants he was raised in Montana, graduated from the University of Montana in Missoula, and went on to represent that state throughout his political career.

Mansfield left home in 1917 before he had completed the 8th grade and joined the United States Navy at the age of fourteen. Ten months of his nineteen months of World War I Navy service were spent overseas, and he subsequently spent one year in the Army. Then, on November 10, 1920, Mansfield enlisted again, this time in the Marine Corps. He served in the Western Recruiting Division at San Francisco until January 1921, when he was transferred to the Marine Barracks at Puget Sound, Washington. The following month he was detached to the Guard Company at Marine Barracks Mare Island in California, and in April boarded the *USAT Sherman* bound for Marine Barracks, Olongapo in the Philippines. One year later Mansfield was assigned to Company A, Marine Battery, Asiatic Fleet, and a short tour of duty with that unit took him along the coast of China before he returned to Olongapo in May of 1922.

That August Marine Private Michael J. Mansfield returned to Cavite in preparation for his return to the United States and eventual discharge, and on November 9, 1922 he was released from active duty and awarded the Good Conduct Medal, with his character being described as "excellent" during his two years as a Marine.

Mansfield returned to Montana after his discharge and worked in the Butte mines as a miner and mining engineer until 1930. Having never attended high school, he had to learn to read in order to study for the entrance examinations required to enter college. He then attended the Montana School of Mines from 1927 to 1928, and the University of Montana from 1930 to 1934, where he was awarded B.A. and M.A. degrees and went on to teach for ten years. Before being elected to his first term in Congress in 1942, Mansfield was the Professor of Latin American and Far Eastern History at the University of Montana and a member of the American Federation of Teachers. He then served as a member of the Democratic Party in the U.S. House of Representatives from 1943 until 1953, and in the United States Senate from 1953 until 1977.

An early supporter of Vietnamese President Ngo Dinh Diem, Mansfield had a change of heart after a visit to Vietnam in 1962. He reported to President Kennedy that much of the money given to Diem's government was being squandered, and that the U.S. should avoid further involvement in Vietnam - thus becoming the first American official to comment adversely on the war's progress.

During the subsequent Johnson Presidency Mansfield became a frequent and vocal critic of U.S. involvement in the war. He later hailed the new Nixon administration, especially the "Nixon Doctrine" which was announced on Guam in 1969 and stated that the U.S. would honor all U.S. treaty commitments against those who might invade the lands of allies of the United States, would provide a nuclear umbrella against threats of other nuclear powers, and would supply weapons and technical assistance to countries where warranted - all without committing American forces to local conflicts.

In turn Nixon looked to Mansfield for advice and tapped him as his liaison with the Senate on matters pertaining to Vietnam. Nixon began a steady withdrawal of U.S. troops shortly after taking office in January of 1969, a policy supported by Mansfield, and during his first term reduced American forces by ninety-five percent. The last ones left in March of 1973.

During the economic crisis of 1971 Mansfield was not afraid to reach across the aisle to help the economy. He said, "What we're in is not a Republican recession, or a Democratic recession. Both parties had much to do with bringing us where we are today, but we're facing a national situation which calls for the best all of us can produce, because we know the results will be something which we will all regret."

Mansfield retired from the Senate in 1976 and was appointed Ambassador to Japan in April of 1977 by Jimmy Carter, a role he retained during the Reagan administration through 1988. While serving there Mansfield and his wife were highly respected, and he is still a household name in Japan to this day. He is particularly renowned for describing the United States-Japan relationship as the "most important bilateral relationship in the world, bar none," and his successor noted in his memoirs that the phrase was a "mantra" for Mansfield.

Mansfield retired in 1989 and has been honored many times since. In that year he received the Presidential Medal of Freedom, as well as the United States Military Academy's Sylvanus Thayer Award. In 1990 Japan conferred on Ambassador Mansfield the Grand Cordon of the Order of the Rising Sun with Paulownia Flowers, which is Japan's highest honor for someone who is not a head of state.

The Maureen and Mike Mansfield Memorial Library at the University of Montana in Missoula is named after both him and his wife Maureen, as was his request when informed of the honor. The library contains the Maureen and Mike Mansfield Center, which is dedicated to Asian studies and "advancing understanding and co-operation in U.S.-Asia relations." The Mike Mansfield Federal Building and United States Courthouse in Missoula were also renamed in his honor in 2002.

Mike Mansfield died from congestive heart failure at the age of ninety-eight on October 5, 2001, and was the last known veteran of the "Great War" to die before reaching the age of one hundred. The burial plot of Senator and Mrs. Mansfield can be found in section 2, marker 49-69F of Arlington National Cemetery, which is significant in light of the following remarks by Marine Colonel James M. Lowe on October 20, 2004.

"This gentleman went from snuffy to national and international prominence, and when he died in 2001 he was rightly buried in Arlington. If you want to visit his grave, don't look for him near the Kennedy Eternal Flame, where so many politicians are laid to rest. Look for a small, common marker shared by the majority of our heroes. Look for the marker that says, 'Michael J. Mansfield, PFC, U.S. Marine Corps.'"

PETE MCCLOSKEY
Congressman from California

Paul Norton "Pete" McCloskey Jr. is a former Republican politician from California who served in the U.S. House of Representatives from 1967 to 1983. He ran for the Republican nomination for President on an anti-war platform in 1972, but was defeated by incumbent President Richard Nixon, and in April of 2007 switched his affiliation to the Democratic Party. He is a highly decorated Marine Corps veteran of the Korean War, and has been awarded the Navy Cross, the Silver Star, and two Purple Hearts. His fourth book, *The Taking of Hill 610,* describes some of his exploits in Korea.

McCloskey was born on September 29, 1927 in Loma Linda, and attended public schools in South Pasadena and San Marino. He then attended Occidental College and the California Institute of Technology under the Navy's V-5 Pilot Program, and graduated from Stanford University in 1950 and Stanford University Law School in 1953.

McCloskey voluntarily served in the Navy from 1945 to 1947, the Marine Corps from 1950 to 1952, the Marine Corps Reserve from 1952 to 1960, the Ready Reserve from 1960 to 1967, and retired from the Marine Corps Reserve in 1974 as a Colonel.

He was Deputy District Attorney for Alameda County, California from 1953 to 1954, practiced law in Palo Alto from 1955 to 1967, and was a lecturer on legal ethics at the

Santa Clara and Stanford Law Schools from 1964 to 1967. McCloskey was first elected as a Republican to the 90th Congress in a special election to fill the vacancy caused by the death of U.S. Rep. J. Arthur Younger, was reelected to the seven succeeding Congresses, and served from December 12, 1967 to January 3, 1983.

McCloskey sought the 1972 Republican Presidential nomination on a pro-peace/anti-Vietnam War platform and received one vote (out of 1324) from a New Mexico delegate. All other votes cast went to President Nixon, and thus McCloskey finished in second place.

McCloskey was not a candidate for reelection in 1982, but was instead an unsuccessful Republican candidate for the United States Senate. The 1982 California Republican Senatorial primary was a contentious battle between McCloskey, Maureen Reagan (daughter of then-President Ronald Reagan), Representative Barry Goldwater Jr. (son of Arizona Senator and 1964 Republican Presidential nominee Barry Goldwater), and San Diego Mayor (and fellow Marine) Pete Wilson, who was the eventual victor.

McCloskey was the first member of Congress to publicly call for the impeachment of President Nixon after the Watergate scandal and Saturday Night Massacre, and was also the first to call for a repeal of the Gulf of Tonkin Resolution which had allowed the U.S. to enter the War in Vietnam.

In the late 1980s religious broadcaster (and former Marine) Pat Robertson sued Congressman McCloskey and Representative Andrew Jacobs, Jr. for libel. McCloskey had made remarks, which Jacobs later repeated, alleging Robertson had used connections to avoid combat duty in the Korean War. Eventually Robertson dropped the lawsuit, claiming scheduling conflicts between court dates and his

1988 presidential campaign, and was ordered to pay part of McCloskey's court costs.

In January of 2006 McCloskey announced that he would return to the political arena by running against seven-term incumbent Republican Richard Pombo in the Republican primary for California's 11th Congressional District. Earlier in the year he had formed a group called the "Revolt of the Elders" to recruit a viable primary candidate to run against Pombo. McCloskey's aging campaign bus sported the slogan "Restore Ethics to Congress," and he commented, "Congressmen are like diapers. You need to change them often, and for the same reason." Although he was endorsed in the primary by the *San Francisco Chronicle* and *Los Angeles Times* McCloskey was defeated by Pombo and subsequently backed Jerry McNerney, a Democrat who would go on to unseat Pombo in the 2006 midterm elections. The Sierra Club later recognized McCloskey for helping to unseat the anti-environmentalist Pombo with their 2006 Edgar Wayburn Award.

In the spring of 2007 McCloskey announced that he had changed his party affiliation to the Democratic Party, and in a letter to the *Tracy Press* stressed that the "new brand of Republicanism" had finally led him to abandon the party he had joined in 1948. He followed up with an op-ed column in which he explained that "Disagreement (with party leadership) had turned into disgust" and "I finally concluded it was fraud to remain a member of this modern Republican Party," although it was a "decision not easily taken."

Pete McCloskey was co-chair of the first Earth Day in 1970, is pro-choice, supports stem cell research and assisted suicide, and in 1972 published a book called *Truth and Untruth: Political Deceit in America*.

133

PAUL F. MCHALE, JR.

Congressman from Pennsylvania
Assistant Secretary of Defense

Paul F. McHale, Jr. was the Assistant Secretary of Defense for Homeland Defense and was the member of Congress representing Pennsylvania's 15th Congressional District in the United States House of Representatives from 1993 to 1999.

McHale was born on July 26th, 1950 in Bethlehem, Pennsylvania, where he later graduated from Liberty High School. He received a Bachelor of Arts degree from Lehigh University in 1972, a J.D. from Georgetown University Law Center in 1977, and served in the Marine Corps from 1972 to 1974. He has been a member of the Marine Corps Reserve since 1974, with the rank of Colonel, and served in Operations Desert Shield and Desert Storm during the Gulf War.

McHale was a member of the Pennsylvania House of Representatives from 1983 to 1991, but resigned in order to volunteer for active duty in the Gulf War. Then in 1992 he ran for Congress and defeated fourteen-year incumbent Don Ritter, and was later reelected twice.

McHale gained prominence in 1998 when he called for President Bill Clinton to resign. He voted for three of the four articles of impeachment, was one of the few Democrats to support the measure, and had by far the most liberal voting record of those who were in favor of removing

Clinton from office (the other four Democrats who voted for at least one article were Virgil Goode, Ralph Hall, Charlie Stenholm, and Gene Taylor, all of whom had very conservative voting records. Two of them, Goode and Hall, subsequently became Republicans).

McHale assumed his position as the Assistant Secretary of Defense for Homeland Defense on February 7, 2003 and served in that capacity until January of 2009. During that time, in October of 2006, he was once again recalled to active duty by the Marine Corps to deploy to Afghanistan.

As of this date Paul McHale is engaged to Martha Rainville, a former Congressional candidate from Vermont.

SID MCMATH
Governor of Arkansas

Sidney Sanders McMath (June 14, 1912 - Oct 4, 2003) was an attorney, a decorated Marine, and the 34th Governor of Arkansas. In defiance of his state's political establishment he championed rural electrification, massive highway and school construction, the building of the University of Arkansas for Medical Sciences, strict bank and utility regulation, repeal of the poll tax, open and honest elections, and the broad expansion of opportunities for black citizens in the decade following World War II.

McMath remained loyal to President Harry S. Truman during the "Dixiecrat" rebellion of 1948, campaigning throughout the South for Truman's re-election. As a former Governor, McMath led the opposition to segregationist Governor Orval Faubus following the 1957 Little Rock school crisis. He later became one of the nation's foremost trial lawyers, representing thousands of injured persons in precedent-setting cases and mentoring several generations of young attorneys.

McMath was born in a log cabin on the old McMath home place near Magnolia, Arkansas, the son of Hal Pierce and Nettie Belle Sanders McMath. His paternal grandfather, Columbia County Sheriff Sidney Smith McMath, grand nephew of his Goliad namesake, had been killed in the line of duty the previous year, leaving a pensionless widow and eight children, with Hal being the eldest. After years of wrangling horses and bad-luck wildcatting in the Southwest Arkansas oil fields Hal McMath moved his family by wagon to Hot Springs in June of 1922, where he sold the last of his horses and took a job as a barber.

Sid and his sister Edyth attended Hot Springs public schools, where the boy excelled in boxing and drama and became an Eagle Scout, while shining shoes and hawking newspapers to supplement the family's meager income. He was elected president of his class in each of his high school years, and won the state Golden Gloves welterweight boxing title as a senior. He attended Henderson State College and the University of Arkansas, where he was elected president of the student body, and graduated from the University's School of Law in 1936.

McMath received a reserve ROTC commission as a Second Lieutenant in the Marine Corps upon graduation from college, and during World War II served with the Marines after voluntarily returning to active duty in 1940. After first being assigned to train officer candidates at Quantico he was promoted to captain, and then to major, and in 1942 McMath was ordered to American Samoa where he assumed command of the combined forces jungle warfare school. Then from late 1942 to early 1944 he led the 3rd Marine Regiment in battle as operations officer and acting CO on New Georgia, Vella Lavella, Guadalcanal and Bougainville. During the latter campaign he directed the Battle of Piva Forks, which was the pivotal action, and single-handedly rallied Marines who were pinned down by enemy mortar and machinegun fire. Afterwards he received a battlefield promotion to Lieutenant Colonel and was awarded the Silver Star and Legion of Merit. The citation for the former, which was personally signed by Admiral W.F. "Bull" Halsey, lauded McMath's "extraordinary heroism...

137

and disregard for his own safety above and beyond the call of duty (which) was an inspiration to the officers and men who observed him." Shortly afterward McMath was stricken with malaria and filariasis, and was hospitalized for several months in San Diego, California. He then served at the Marine Corps' headquarters in Washington, D.C. and contributed to planning an amphibious invasion of the Japanese home islands before being discharged from active duty in December of 1945 as a Lieutenant Colonel.

McMath resumed his service in the Marine Corps Reserve following his tenure as Governor, and commanded VTU 8-14 in Little Rock until 1964. He also held the office of National President of the 3d Marine Division Association from 1960 to 1961.

Following a promotion to Brigadier General in June of 1963 McMath returned to active duty as the Assistant Commanding General, Marine Corps Base Camp Pendleton during the summer of 1963. He subsequently served as Assistant Commanding General, Landing Force Training Unit Pacific at Coronado in the summer of 1964, Assistant Division Commander, 2nd Marine Division at Camp Lejeune in the summer of 1965, and President of the Marine Corps Reserve Policy Board in the summer of 1966. In addition McMath served with the 3rd Marine Amphibious Force in Vietnam, was promoted to Major General in November of 1966, and became Assistant Deputy Commander, Fleet Marine Force, Atlantic in 1967, and served a second brief Reserve tour in Vietnam with the 3rd Marine Division in 1969.

In 1967 McMath helped found the Marine Corps JROTC at the Catholic High School for Boys in Little Rock, Arkansas, which went on to become one of the top JROTC units in the country - and ever since, the cadets there have been known as "Sid's Kids."

In early 1946 McMath and other veterans returning to Arkansas from World War II banded together to fight corruption in the Hot Springs city government, which was dominated by illegal gambling interests. At the time Hot Springs was a national gambling mecca frequented by organized crime figures from Chicago, New York, and other metropolitan areas. Mobsters maintained political control by purchasing and holding hundreds of poll tax receipts, often in the names of deceased or fictitious persons, which would be used to cast multiple votes in different precincts, and law enforcement officers were on the payroll of the local "organization" headed by long-serving Mayor Leo McLaughlin. McMath headed a "GI Ticket" which, except for McMath himself, was defeated in the Democratic primary election, so the others resigned from the party and ran again as independents in the 1946 general election after McMath persuaded a federal judge to toss out the fraudulent poll tax receipts. Most won their offices, among them noted combat aviators Earl Ricks and I.G. Brown, who were elected mayor and sheriff.

McMath served as prosecuting attorney for the 18th Judicial District starting in 1947, and the newly installed GI officials, led by McMath, shut down the casinos and other rackets. A grand jury indicted a number of owners, pitchmen and politicians, including the former Mayor, and with the coming development of Las Vegas, Hot Springs lost its premier gaming status.

After his success as a prosecutor, McMath was elected Governor in 1948 in a close election. He entered office in January of 1949 as the nation's youngest Governor, and was easily reelected in 1950 over his immediate predecessor, Ben

Laney, who attacked McMath for supporting Truman in 1948 when Laney and a number of other southern Democrats bolted the party over its civil rights plank. The walk-outs switched their allegiance to Governor Strom Thurmond of South Carolina, who ran as a "Dixiecrat," and as McMath campaigned vigorously across the region he was credited by Truman with helping to save most of the South for the Democratic column. The two developed a lifelong friendship, and McMath was mentioned as a possible Vice-Presidential choice in 1952.

McMath's administration focused on infrastructure improvements, including the extensive paving of farm-to-market and primary roads "to get Arkansas out of the mud and the dust," rural electrification, and the construction of a medical center in the capitol city. McMath supported anti-lynching statutes, and appointed African Americans to state boards for the first time. His administration consolidated hundreds of small school districts, built the University of Arkansas for Medical Sciences, and worked tirelessly to save the state's all-black college, Arkansas Agricultural, Mechanical, & Normal (now the University of Arkansas at Pine Bluff).

McMath eventually ran afoul of the energy companies as well as other sectors which had long dominated Arkansas politics - but for whom McMath was not a compliant agent. These included Arkansas Power & Light, wealthy bankers and bond dealers, piney woods timber companies, the Murphy Oil conglomerate, and old-family planters in the Mississippi Delta. All feared McMath's progressive politics would increase labor costs and break up the sharecropping farm economy. These interests put aside their differences to work in concert to defeat McMath's bid for a third term in the 1952 election, and he ran unsuccessfully for the U.S.

Senate in 1954 and again for Governor in 1962 with largely the same opposition. McMath's voting base among the working class was neutralized by the two dollar poll tax which had to be paid a year prior to an election and effectively disenfranchised thousands of voters. He had tried to repeal the tax, but it remained a relic of Jim Crow until the 24th Amendment to the U.S. Constitution in 1964.

After his 1952 defeat McMath returned to the practice of law and over the next half-century became one of the leading consumer trial attorneys in the United States. He and his partner Henry Woods, who had served as his Gubernatorial Chief of Staff, became nationally known for their effective use of powerful demonstrative evidence such as detailed models of accident scenes and the human anatomy.

In a 1999 opinion poll of political science professors McMath placed fourth on a list of top Arkansas Governors of the 20th century, however in a December 2003 forum of historians and journalists sponsored by the Old State House Museum in Little Rock there was a consensus that his achievements could well result in his elevation by future historians to first place - not only among Arkansas Governors, but among all Southern Governors of the time.

"Sid McMath might have laid legitimate claim to have been the most courageous and far-sighted Southern leader of the 20th century," wrote *Arkansas Times* columnist Ernest Dumas on October 10, 2003. "What separated McMath from every other leader of that grim time in the South was courage, the moral as well as physical variety."

The *Arkansas Democrat-Gazette*, in an October 7, 2003 editorial called *Greatness Passed This Way* (written by editorial page editor Paul Greenberg, himself a recipient of the Pulitzer Prize) lauded McMath as "the greatest (man) of his era - and of a few others."

Sid McMath served a bare six years in public office, only four as Governor, and left behind no powerful political organization or claque of partisans. Whatever fame McMath once had fled well before his death. In recent years he sometimes had to spell his name for bank tellers, reservations clerks, state employees - once even for a newspaper reporter. He accumulated no great wealth, owning at the end a modestly upscale condominium and a small residual interest in his law firm. The latter, though no longer occupying the field alone, remains the state's premier personal injury practice.

Sid McMath died at his home in Little Rock on Saturday, October 4, 2003 at the age of ninety-one. He was survived by his third wife, Betty Dorch Russell McMath, three sons, and two daughters. His first wife and childhood sweetheart, Elaine Braughton McMath, died at Quantico, Virginia in 1942, and his second wife of forty-nine years, Anne Phillips McMath, died at Little Rock in 1994.

McMath was given a full military funeral by a Marine Corps Honor Guard, lay in state for a day in the Capitol rotunda, and was eulogized by Former Governor David Pryor as, "the best friend Arkansas ever had." Following the firing of a salute by the Honor Guard, McMath was interred at Pinecrest Memorial Cemetery in Saline County, Arkansas - fittingly, just a few yards from a survey marker denoting the geographical center of the state.

Sid McMath Avenue in Little Rock is named for him, and in December of 2004 the Central Arkansas Library System dedicated a new branch in his honor. A statue of McMath waving his trademark Panama campaign hat was commissioned by the library, and was unveiled in September of 2006 as the centerpiece of a sculpture plaza and nature trail.

McMath wrote a memoir, *Promises Kept*, detailing his rural upbringing, public schooling, and family tragedies - including the untimely death of his first wife Elaine during the war, and the shooting to death of his father, who had become an enraged alcoholic, by his second wife Anne in 1947 - as well as his years of service in the military and as Governor. The Arkansas Historical Association awarded the autobiography its 2003 John G. Ragsdale Prize as the year's most outstanding historical work, and in April of 2006 the book was awarded the Booker T. Worthen Medal for literary excellence by the Arkansas Library System.

ZELL MILLER
Governor of and Senator from Georgia

Zell Bryan Miller is a politician from the state of Georgia. Miller served as Lieutenant Governor from 1975 to 1991, the 79th Governor of Georgia from 1991 to 1999, and as a United States Senator from 2000 to 2005.

Although a Democrat, Miller famously backed Republican President George W. Bush over Democratic nominee John Kerry in the 2004 presidential election, and since 2003 has frequently criticized the Democratic Party and publicly supported several Republican candidates. During the 2004 election Miller gained significant public exposure after he became visibly angry during an interview on *Hardball with Chris Matthews,* and in 2006 he did voice-overs (narrations) for Republican candidate commercials in Georgia state elections.

Zell Miller was born on February 24, 1932 in the small mountain town of Young Harris, Georgia. His father died when Miller was an infant, and the future politician was raised by his widowed mother. As a child he lived in both Young Harris and Atlanta, and today Miller lives in the old Young Harris family home. He spent his first two years of college at Young Harris College in his hometown, and holds Bachelor's and Master's degrees in history from the University of Georgia.

Less than a month after the Korean War ended Miller wound up in a drunk tank in the North Georgia Mountains, and later claimed this incident was the lowest point of his life. Upon his release Miller enlisted in the Marine Corps, and during his three years in the Corps attained the rank of Sergeant. He often refers to the value of his experiences in the Marine Corps in his writing and stump speeches, and in his book on the subject, entitled *Corps Values: Everything You Need to Know I Learned in the Marines*, he wrote, "In the twelve weeks of hell and transformation that were Marine Corps boot camp, I learned the values of achieving a successful life that have guided and sustained me on the course which, although sometimes checkered and detoured, I have followed ever since."

Miller's parents were both involved in local politics in the North Georgia mountains. His father, a Democrat, was Mayor of Young Harris from 1959 to 1960, was elected to two terms as a Georgia State Senator during the 1960s, and in 1964 and 1966 unsuccessfully sought the Democratic nomination for a seat in the United States House of Representatives. He endorsed segregation in both races, and later served in several positions in state government and the Georgia Democratic Party.

Miller's first experience in the executive branch of government was as Chief of Staff for Georgia Governor Lester Maddox. He was elected Lieutenant Governor of Georgia in 1974, and served four terms from 1975 to 1991 during the terms of Governors George Busbee and Joe Frank Harris, making him the longest-serving Lieutenant Governor in Georgia history. Then in 1980 he unsuccessfully challenged Herman Talmadge in the Democratic primary for his seat in the United States Senate.

Miller was elected Governor of Georgia in 1990, defeating Republican Johnny Isakson after besting Atlanta Mayor Andrew Young and future Governor Roy Barnes in the primary. Miller campaigned on the concept of term limits and pledged to seek only a single term as Governor - although he later ran for and won reelection, with fellow Marine James Carville serving as his campaign manager. In 1991 Miller endorsed Governor Bill Clinton of Arkansas for President of the United States. He became close to Clinton, and some political commentators described Miller's support as critical in helping Clinton hold the South and secure the nomination after a rocky start in the Democratic primaries. Miller gave the keynote speech at the 1992 Democratic National Convention at Madison Square Garden in New York City, and in two oft-recalled lines said that President George H. W. Bush "just doesn't get it," and remarked of a statement by Vice President Dan Quayle, "I know what Dan Quayle means when he says it's best for children to have two parents. You bet it is! And it would be nice for them to have trust funds, too. We can't all be born rich and handsome and lucky. And that's why we have a Democratic Party. My family would still be isolated and destitute if we had not had FDR's Democratic brand of government. I made it because Franklin Delano Roosevelt energized this nation. I made it because Harry Truman fought for working families like mine. I made it because John Kennedy's rising tide lifted even our tiny boat. I made it because Lyndon Johnson showed America that people who were born poor didn't have to die poor. And I made it because a man with whom I served in the Georgia Senate, a man named Jimmy Carter, brought honesty and decency and integrity to public service."

Upon leaving the Governor's office in January of 1999, Miller accepted teaching positions at Young Harris College, Emory University and the University of Georgia, and was a visiting professor at all three institutions when he was appointed to the U.S. Senate. His successor as Governor, Roy Barnes, appointed Miller to a U.S. Senate seat following the death of Republican Senator Paul Coverdell in July 2000. While the Democratic Party's historic control of Georgia politics had waned for years, Miller remained popular and easily won a special election to keep the seat in November of 2000. During the campaign he spoke warmly of his late friend Coverdell, praised Republican presidential candidate George W. Bush, and promised to work for bipartisanship in the Senate. As Coverdell had last been elected in 1998, Miller had four years remaining in the Senate term before his retirement from politics in January of 2005.

Throughout Zell Miller's career as a U.S. Senator he showed increasing support for Republicans and increasing criticism of Democrats, leading some to question whether his fellow Democrats in the Senate had given him a lukewarm reception. However, given his beginnings as a conservative southern Democrat, it is likely he found his views drastically different from the more liberal ideology of the national party.

During 2001 and 2002, when liberal Republican Senators from New England like James Jeffords and Lincoln Chafee threatened to (and in Jeffords' case, did) leave their party over ideological disputes, rumors abounded that Miller would become a Republican in order to return control of the Senate to that party. These rumors were dispelled with Miller saying, "I'll be a Democrat 'til the day I die."

In 2003 Miller announced that he would not seek re-election after completing his term in the Senate. He also announced he would support President George W. Bush in the 2004 presidential election rather than any of the nine candidates then competing for his own party's nomination. He maintained this position after fellow Senator John Kerry became the Democratic nominee, and Miller, who had been the keynote speaker at the 1992 Democratic National Convention, was subsequently announced as keynote speaker at the 2004 Republican National Convention.

In 2004 he cosponsored the Federal Marriage Amendment to the United States Constitution. If it had been ratified, it would have declared that marriage in the United States only consists of the union of a man and a woman, and would have prohibited state and the federal government from recognizing same-sex marriages and same-sex domestic partnerships. In March of that year he introduced the Broadcast Decency Responsibility and Enforcement Act, which would have created a Council of Decency to advise the Federal Communications Commission on standards of decency in broadcasting. The Council would have consisted of three individuals from the ministry, three broadcast industry representatives, and three school teachers. The money from penalties for obscene, indecent, and profane broadcasts would have been given to faith-based organizations.

Miller established himself as a conservative on virtually all economic issues. He was the first Democrat in the Senate to publicly declare his support for the Economic Growth and Tax Relief Reconciliation Act of 2001, a broad-based tax cut which was criticized by opponents for favoring the rich and being fiscally irresponsible. Miller was the only Democrat to vote against an amendment to that same bill which was submitted by Tom Harkin (D-Iowa) to scale back portions of the tax cut in order to spend more on education and debt reduction. He strongly opposed the estate tax, and voted a number of times for its repeal. He also advocated drilling in the Arctic National Wildlife Refuge.

Miller argued in his book *A National Party No More: The Conscience of a Conservative Democrat* (authored and published in 2003) that the Democratic Party lost its majority because it does not stand for the same ideals that it did in the era of John F. Kennedy. He argued that the Party, as it now stands, is a far left-wing organization that is out of touch with the America of today and that the Republican Party now embraces the conservative Democratic ideals he has held for so long. The book spent nine weeks on *the New York Times* Best Seller list for hardback non-fiction.

Despite Miller's frequent disagreements with his own party, he did occasionally support some of their positions. For example, he was a strong supporter of the Bipartisan Campaign Reform Act of 2002. In Miller's view the provisions of the bill, limiting donations to candidates for political office, should have gone even further. Miller voted with all his fellow Democratic Senators for the Bipartisan Patient Protection Act, and later voted with virtually all Democrats to allow American consumers to import cheaper prescription drugs from Canada - a bill which was strongly opposed by American pharmaceutical companies.

In his keynote convention speech, delivered in September of 2004, Miller criticized the current state of the Democratic Party. He said, "No pair has been more wrong, more loudly, more often than the two Senators from Massachusetts - Ted Kennedy and John Kerry." He also criticized Kerry's Senate voting record, claiming that his votes against bills for defense and weapon systems indicated support for weakening U.S. military strength, saying "the B-1 bomber that Senator Kerry opposed dropped forty percent of the bombs in the first six months of Enduring Freedom. The B-2 bomber that Senator Kerry opposed delivered air strikes against the Taliban in Afghanistan and Hussein's command post in Iraq. The F-14A Tomcats that Senator Kerry opposed shot down Khadafi's Libyan MIGs over the Gulf of Sidra. The modernized F-14D that Senator Kerry opposed delivered missile strikes against Tora Bora. The Apache helicopter that Senator Kerry opposed took out those Republican Guard tanks in Kuwait in the Gulf War. The F-15 Eagles that Senator Kerry opposed flew cover over our Nation's Capital and this very city after 9/11. I could go on and on and on - against the Patriot Missile that shot down Saddam Hussein's scud missiles over Israel, against the Aegis air-defense cruiser, against the Strategic Defense Initiative, against the Trident missile, against, against, against. This is the man who wants to be the Commander in Chief of the U.S. Armed Forces? U.S. forces armed with *what?* Spitballs?"

The speech was well received by the convention attendees, especially the Georgia delegates. Conservative commentator Michael Barone compared the remarks to the views and ideology of Andrew Jackson. Miller's combative reaction to post-speech media interviews received almost as much attention as the speech itself. First, in an interview with CNN, Miller had a dispute with Judy Woodruff, Wolf Blitzer, and Jeff Greenfield when they questioned him on his speech, particularly on whether he had misinterpreted the context and full content of Kerry's votes, and the fact that Dick Cheney, as Defense Secretary, had opposed some of the same programs he had attacked Kerry for voting against.

Shortly thereafter, Miller appeared in an interview with Chris Matthews on the MSNBC show *Hardball*. Here, Miller became visibly angry. Matthews criticized the premise of Miller's assertion that Kerry had actually voted against such defense programs by noting that in voting on appropriations bills, Senators often vote against a version of a bill without wishing to oppose every item in that bill. Matthews also asked Miller to compare his assertion that a military under Kerry would be armed with only "spitballs" with rhetoric from Democrats that Republicans "want to starve little kids, they want to get rid of education, they want to kill the old people" and whether such level of rhetoric was constructive. When Miller expressed irritation at this line of questioning, Matthews pressed Miller with the question, "Do you believe now - do you believe, Senator, truthfully, that John Kerry wants to defend the country with spitballs?" Miller at first said that he wished the interview had been face-to-face so that he could "get a little closer up into your face." He then angrily told Matthews to "get out of my face" and declared, "I wish we lived in the day where you could challenge a person to a duel." The interview was later parodied on *The Daily Show with Jon Stewart*, *Late Night with Conan O'Brien*, and *Saturday Night Live*.

After President Bush was re-elected Miller referred to the Republican victories in that election as a sign that Democrats didn't relate to most Americans. Calling for Democrats to change their message, he authored a column which appeared in the *Washington Times* in November of 2004 in which he wrote, "Fiscal responsibility is unbelievable in the face of massive new spending promises. A foreign policy based on the strength of 'allies' like France is unacceptable... a strong national defense policy is just not believable coming from a candidate who built a career as an anti-war veteran, an anti-military candidate, and an anti-action Senator. When will national Democrats sober up and admit that dog won't hunt? Secular socialism, heavy taxes, big spending, weak defense, limitless lawsuits and heavy regulation - that pack of beagles hasn't caught a rabbit in the South or Midwest in years."

Zell Miller is lauded in conservative circles but is increasingly distant from the Democratic Party. He declared early in 2008 that he would not support either Senator Barack Obama or Senator Hillary Clinton in the presidential election, and after Obama was elected President and the

Democrats increased their majorities in the House and Senate he endorsed Republican Saxby Chambliss in the Senate run-off against Democrat Jim Martin and criticized Obama over "spreading the wealth."

After leaving the Senate Zell Miller joined the law firm McKenna Long & Aldridge in the firm's national Government Affairs practice. He is also a frequent *Fox News Channel* contributor, and currently serves on the Board of Directors of the National Rifle Association.

The former Student Learning Center at the University of Georgia was renamed the 'Miller Student Learning Center' in October of 2008.

JOHN MURTHA
Congressman from Pennsylvania

John Patrick "Jack" Murtha, Jr. (June 17, 1932 - Feb 8, 2010) was a Democrat politician from Pennsylvania who represented that State's 12th Congressional District in the United States House of Representatives from 1974 until his death in 2010.

A former Marine Corps officer, Murtha was the first Vietnam War veteran elected to the U.S. House of Representatives. Previously a member of the Pennsylvania House from 1969 to 1974, in 1974 he narrowly won the special election held to choose a successor when the incumbent died in office. In the first decade of the 21st century Murtha has been best known for calling for the withdrawal of American forces in Iraq, as well as questions about his ethics.

Murtha was born into an Irish-American family in New Martinsville, West Virginia near the border of Ohio and Pennsylvania and grew up in a largely suburban county east of Pittsburgh. As a youth he became an Eagle Scout, and worked at a gas station and delivered newspapers before graduating from the Kiski School, an all-male boarding school in Saltsburg, Pennsylvania.

Murtha left Washington and Jefferson College in 1952 to join the Marine Corps and was awarded the American Spirit Honor Medal for displaying outstanding leadership qualities during training. He became a drill instructor at Parris Island, was selected for Officer Candidate School at Quantico, and was then assigned to the Second Marine Division at Camp Lejeune in North Carolina.

Murtha left the Marines in 1955, but remained in the Reserves until volunteering for service in the Vietnam War. He served there from 1966 to 1967 as a battalion staff officer (S-2 Intelligence) and received the Bronze Star with Valor device, two Purple Hearts, and the Vietnamese Cross of Gallantry. Murtha retired from the Marine Corps Reserve as a Colonel in 1990 and received the Navy Distinguished Service Medal.

Murtha was elected to represent the 72nd Legislative District in the Pennsylvania House of Representatives in May of 1969 and to Congress in 1974. He faced tough primary challenges in 1982, 1990 and again in 2002, with the 1982 challenge occurring when the Republican-controlled State Legislature took advantage of Murtha's connection to the 'Abscam' scandal and incorporated most of the district of fellow Democrat and Vietnam War veteran Don Bailey into the 12th District.

The 2002 challenge occurred when the State Legislature redrew the district of fellow Democrat Frank Mascara to make it more Republican-friendly, and shifted a large chunk of Mascara's former territory into Murtha's district. Mascara opted to run against Murtha in the Democratic primary since the new 12th contained more of his old territory than Murtha's, but was badly defeated.

In 2006 Murtha's Republican challenger was Diana Irey, a County Commissioner from the heart of Mascara's former district. Irey attacked Murtha for his criticism of the Iraq war, and even though she was Murtha's strongest Republican opponent in decades she polled well behind him throughout the campaign and lost in a landslide.

In 2006, after the Democrats won control of Congress in the 2006 midterm elections, Murtha made a failed bid to be elected House Majority Leader for the 110th Congress with the open support of new House Speaker Nancy Pelosi. He lost to Steny Hoyer of Maryland, and after this defeat became Chairman of the House Appropriations Defense Subcommittee, which he had previously chaired from 1989 to 1995 and served on as ranking Democrat from 1995 to 2007.

In 1980, during his fourth term in Congress, Murtha became embroiled in the Abscam investigation, which targeted dozens of Congressmen. The investigation centered on FBI operatives posing as intermediaries for Saudi nationals hoping to bribe their way through the immigration process and into the United States. Murtha met with these operatives and was videotaped, and later agreed to testify against Frank Thompson (D-NJ) and John Murphy (D-NY), the two Congressmen mentioned as participants in the deal at the same meeting who were later videotaped placing cash bribes in their trousers. The FBI videotaped Murtha responding to an offer of $50,000, with Murtha saying, "I'm not interested... at this point. (If) we do business for a while, maybe I'll be interested, maybe I won't" - right after Murtha offered to provide the names of businesses and banks in his district where money could be invested legally.

Murtha was targeted by Citizens for Responsibility and Ethics in Washington (CREW) as one of the twenty most corrupt members of Congress, and in September of 2006 that organization listed him under 'Five Members to Watch' in its Second Annual Most Corrupt Members of Congress Report. The report cited Murtha's steering of defense appropriations to clients of KSA Consulting, which employed his brother Robert, and to the PMA Group, founded by former

Appropriations Committee Subcommittee on Defense senior staffer Paul Magliocchetti.

In 2008 *Esquire* magazine named him one of the ten worst members of Congress because of his opposition to ethics reform, and the one hundred million a year he brought to his district in earmarks. The *Wall Street Journal* has called Murtha "one of Congress' most unapologetic earmarkers."

In February of 2009 *CQ Politics* reported that Murtha was one of 104 U.S. Representatives to earmark funds in the 2008 Defense appropriations spending bill for a lobbying group that had contributed to his past election campaigns. The spending bill, which was managed by Murtha in his capacity as Chairman of the House Appropriations Subcommittee on Defense, secured $38.1 million for clients of the PMA Group in the single fiscal law - and they are currently under investigation by the FBI as a result.

Murtha voted for the October 10, 2002 resolution that authorized the use of force against Iraq, but he later began expressing doubts about the war. On March 17, 2004, when Republicans offered a War in Iraq Anniversary Resolution that "affirms that the United States and the world have been made safer with the removal of Saddam Hussein and his regime from power in Iraq" Murtha voted against it.

Still, in early 2005 Murtha argued against the withdrawal of American troops from Iraq. He stated, "A premature withdrawal of our troops based on a political timetable could rapidly devolve into a civil war which would leave America's foreign policy in disarray as countries question not only America's judgment, but also its perseverance."

Then on November 17, 2005 Murtha submitted a resolution in the House of Representatives calling for the redeployment of U.S. troops from Iraq, saying, "The U.S. cannot accomplish anything further in Iraq militarily. It is time to bring them home." The bill cited a lack of progress in stabilizing Iraq, the possibility that a draft would be required to sustain sufficient troop numbers, Iraqi disapproval of U.S. forces, and the increasing costs of the war. The bill proposed that deployment to Iraq be suspended, and that U.S. Marines establish an "over-the-horizon" presence in nearby countries.

Murtha's comments forced a heated debate on the floor of the House, and Republicans led by Duncan Hunter of California, who was Chairman of the House Armed Services Committee, responded by proposing their own resolution which they said was intended to demonstrate that those calling for immediate troop withdrawal from Iraq were "out of the mainstream."

During debate on adopting the rule for the resolution Congresswoman Jean Schmidt (R-Ohio) made a statement attributed to Danny Bubp, an Ohio State Representative and Marine Corps Reservist, saying, "He also asked me to give Congressman Murtha a message. Cowards cut and run, Marines never do."

Seeing Schmidt's remarks as an unwarranted "cheap shot" against Murtha, outraged Democrats brought House business to a halt for ten minutes until Schmidt herself asked for and received permission to withdraw her comments. Bubp has since stated that he never mentioned Murtha when making the quoted comment. He added that he would never question the courage of a fellow Marine, and later said, "I don't want to be interjected into this. I wish (Congresswoman Schmidt) had never used my name."

The Haditha incident occurred the very next day on November 19, 2005, and since then there have been differing accounts of exactly what took place. In November of 2005 Murtha announced that a military investigation into the Haditha killings concluded U.S. Marines had intentionally killed innocent civilians, and in referring to the first report about Haditha that appeared in *Time* magazine said, "It's much worse than reported in *Time* magazine. There was no fire fight. There was no IED that killed these innocent people. Our troops overreacted because of the pressure on them, and they killed innocent civilians in cold blood. And that's what the report is going to tell."

The Marine Corps responded to Murtha's announcement by stating that "there is an ongoing investigation, therefore any comment at this time would be inappropriate and could undermine the investigatory and possible legal process," and Murtha was criticized by conservatives for presenting a version of events as simple fact before an official investigation had been concluded.

In August of 2006 Marine Staff Sergeant Frank Wuterich filed a lawsuit against Murtha for character defamation during an ongoing investigation into the Haditha incident, but in April of 2009 this suit was dismissed by a federal appeals court which ruled Murtha could not be sued because he was acting in his official role as a lawmaker when he made the statements.

On December 21, 2006 the military charged Wuterich with twelve counts of unpremeditated murder against individuals and one count of the murder of six people "while engaged in an act inherently dangerous to others," but charges were subsequently dropped against seven of the eight Marines involved. Only Wuterich is still facing trial, now on nine counts of involuntary manslaughter.

Commenting on the prospects for the election of Barack Obama during the 2008 Presidential campaign, Murtha became the subject of controversy after deriding many of his own constituents as "racists' who would not vote for Obama because he is black. In response to the outrage at his comments he apologized, but then reiterated the point by stating, "There's still folks that have a problem voting for someone because they are black. This whole area, years ago, was really redneck."

Murtha was a Democrat with a relatively populist economic outlook, and was generally much more socially conservative than most other House Democrats. He was pro-life and voted against abortion, however he supported federal funding of embryonic stem cell research. He generally opposed gun control, and was one of the few Democrats in Congress to vote against the Bipartisan Campaign Reform Act of 2002 and in favor of medical malpractice tort reform.

Murtha was strongly pro-labor and opposed both the North American Free Trade Agreement (NAFTA) and the Central American Free Trade Agreement (CAFTA), as well as President George W. Bush's tax plan, Social Security privatization, and the Federal Marriage Amendment. He also he was one of only two Congressmen to vote for a measure proposing reinstating the draft.

Murtha voted for the Affordable Healthcare for America Act, which passed in the House in November of 2009. He said of the bill, "For nearly a century, both Democrats and Republicans have failed to enact comprehensive health care reform. Today's historic vote moves us closer to solving America's health care crisis." Murtha did not support allowing abortions as part of health care reform however, and voted for the Stupak-Pitts Amendment to the bill which prohibits elective abortions for people covered by the public healthcare plan and to prohibit people receiving federal assistance from purchasing a private healthcare plan that includes abortions.

Jack Murtha was hospitalized with gallbladder problems for a few days in December of 2009 and had surgery in

January of 2010 at Bethesda Naval Hospital where "doctors inadvertently cut Mr. Murtha's intestine during the laparoscopic surgery, causing an infection." Due to the complication he was again hospitalized two days later, and died on the afternoon of February 8, 2010 in Arlington, Virginia. Ironically, two days before his death, Murtha had become the longest serving Pennsylvania Congressman in history.

Speaker of the House Nancy Pelosi, who was a close friend of Murtha, said in a statement on the day of his death that "with the passing of John Murtha, America has lost a great patriot," and House Republican Leader John Boehner said "our nation has lost a decorated veteran." On the flip side of the coin, many Marines viewed Murtha as a traitor for condemning the Corps without having any facts, and his passing did not generate the sort of camaraderie normally associated with the death of a high-profile Marine.

CHARLES S. "CHUCK" ROBB
Governor of and Senator from Virginia

Charles Spittal "Chuck" Robb served as the 64th Governor of Virginia from 1982 to 1986, as a United States Senator from 1989 to 2001, and in 2004 chaired the Iraq Intelligence Commission.

Robb was born on June 26, 1939 in Phoenix, Arizona, grew up in the Mount Vernon area of Alexandria, Virginia and earned a Bachelor of Arts degree from the University of Wisconsin–Madison in 1961.

A United States Marine Corps veteran and honor graduate at Quantico, Robb became a White House social aide. It was there that he met and eventually married Lynda Johnson, the daughter of then-President Lyndon B. Johnson. Robb went on to serve two tours of duty in Vietnam, where he commanded a rifle company in combat and was awarded the Bronze Star. After the war he earned a J.D. at the University of Virginia Law School in 1973, and after a clerkship with a federal appeals judge entered private practice.

In 1977 Robb won election as a Democrat for the Lieutenant Governorship of Virginia. He served as Lieutenant Governor from 1978 to 1982, and as Governor from 1982 to 1986. In the 1977 election Robb was the only Democrat running for statewide office to win, making him the head of a political party which had not won a Governor's race in a dozen years. Four years later Robb led the Democrats into office by appealing to conservatives who were disenchanted with his opponent's maverick style.

Virginia Democrats again won all three statewide offices in 1985, which was viewed as an endorsement of Robb's leadership while in office. As a campaigner Robb was capable but reserved, and during a time when political communication styles were beginning to favor sound bites he was known for speaking in paragraphs about complex policy issues. He was also noteworthy among his contemporaries for raising substantial sums of campaign funds.

Politically Robb was a moderate who was known as a conservative Democrat. As Governor he balanced the state budget without raising taxes, and dedicated an additional one billion for education. He appointed a record number of women and minorities to state positions, including the first African American to the State Supreme Court, and was the first Virginia Governor in twenty-five years to use the death penalty. Robb was instrumental in creating the Super Tuesday primary that brought political power to the Southern states, and was also a co-founder of the Democratic Leadership Council.

Robb later served as Democratic member of the U.S. Senate from 1989 until 2001. He ranked annually as one of the most ideologically centrist Senators and often acted as a bridge between Democratic and Republican members, preferring background deal-making to the legislative limelight. His fellow Democrats removed him from the Budget Committee for advocating deeper cuts in federal spending, and in 1991 he was one of a handful of Democratic Senators to support authorizing the use of force to expel Iraqi forces from Kuwait. That same year he was one of only eleven Democrats to vote in favor of Clarence Thomas' controversial nomination to the Supreme Court.

Robb is more liberal on social issues. He voted for the Assault Weapons Ban, against the execution of minors, and was opposed to a Constitutional Amendment to ban flag burning. In 1993 Robb supported President Clinton's proposal to adopt the 'don't ask, don't tell' policy for homosexuals in the armed forces, and three years later he was the only Senator from a Southern state to oppose the Defense of Marriage Act. In stating his opposition to the bill, which his friends and supporters urged him to support, he said, "I feel very strongly that this legislation is wrong. Despite its name, the Defense of Marriage Act does not defend marriage against some imminent, crippling effect. Although we have made huge strides in the struggle against discrimination based on gender, race, and religion, it is more difficult to see beyond our differences regarding sexual orientation. The fact that our hearts don't speak in the same way is not cause or justification to discriminate." Some have speculated that his position on gay rights, along with his positions on other hot-button issues like abortion, only alienated the generally conservative voters of Virginia and contributed to his eventual defeat.

In 1994 Robb narrowly defeated former Iran-Contra figure (and fellow Marine) Oliver North in a poor year nationally for Democrats despite being outspent four to one. Senator John Warner (a former Marine himself) refused to support North, and instead backed third-party candidate and former Virginia Attorney General J. Marshall Coleman, whom Robb had defeated in the 1981 gubernatorial contest. During the campaign, Robb won the endorsement of yet another Marine, former Reagan Naval Secretary (and future U.S. Senator from Virginia) Jim Webb, and high profile Republicans such as Elliot Richardson, William Ruckelshaus, and William Colby. The campaign was later documented in the 1996 film *A Perfect Candidate*.

After his re-election Robb continued to promote fiscal responsibility and a strong national defense, and was the only Senate Democrat to vote for all items in the GOP's "Contract with America" when they reached the Senate floor, including a Balanced Budget Amendment and a line item veto. Robb became the only Senator to simultaneously serve on all three national security committees - Armed Services, Foreign Relations, and Intelligence - but after two terms in the Senate and twenty-five years in statewide politics he was defeated in a close race in 2000 by Republican George Allen, who was also a former Governor.

In 1991 Robb admitted that he had spent time with former Miss Virginia USA Tai Collins alone in a hotel room during the time he was Governor in the 1980s, however he denied having an affair with her - although he admitted to sharing a bottle of champagne and receiving a nude massage. Collins, on the other hand, later told *Playboy* magazine the two had been having an affair since 1983. There were also rumors that during the time he was Governor Robb was present at parties where cocaine was used, but he strongly denied this when the issue was raised during his 1988 campaign for the Senate. In fact Robb so vehemently denied the cocaine allegation that he claimed to not even know what it looked like, which only added to the sensationalism.

Also in 1991, three of Robb's aides resigned after listening to illegally-recorded cell phone conversations of Virginia Governor (and possible 1994 Senate primary opponent) Doug Wilder. The scandal of the phone conversation morphed into a federal grand jury investigation when it was learned Robb's staff as well as Robb himself may have been guilty of conspiring to distribute the contents of an illegal wiretap. Robb and his staff claimed to be unaware that conversations on cell phones are protected by the same laws governing land lines, and the grand jury concluded its work without indicting him.

Following his two terms in the Senate Robb served on the Board of Visitors at the U.S. Naval Academy, and began teaching at George Mason University School of Law. In February of 2004 he was appointed co-chair of the Iraq Intelligence Commission, an independent panel tasked with investigating U.S. intelligence surrounding the United States' 2003 invasion of Iraq and Iraq's weapons of mass destruction, and in 2006 he was appointed to serve on the President's Foreign Intelligence Advisory Board. Robb also served on the Iraq Study Group with former Secretary of State and fellow Marine James A. Baker III, and a *New York Times* article on October 9, 2006 credited him with being the only member of the group to venture outside the American controlled "Green Zone" during a trip to Baghdad. He is also a former member of the Trilateral Commission, and is rumored to be a current member of the Council on Foreign Relations.

Chuck Robb currently resides in McLean, Virginia.

PAT ROBERTS
Senator from Kansas

Charles Patrick "Pat" Roberts is the junior United States Senator from Kansas. A member of the Republican Party, he was formerly the Chairman of the Senate Intelligence Committee.

Roberts was born in Topeka, Kansas on April 20th, 1936 to Ruth B. Patrick and C. Wesley Roberts. His father served for four months as Chairman of the Republican National Committee under Dwight D. Eisenhower, and Roberts' great-grandfather, J.W. Roberts, was the founder of the *Oskaloosa Independent*, which claims to be the second-oldest newspaper in Kansas.

Roberts graduated from high school in 1954, went on to earn a B.A. in Journalism from Kansas State University in 1958, and from 1958 to 1962 served as a Captain in the Marine Corps. He then worked as a reporter and editor for several Arizona newspapers before joining the staff of Republican Kansas Senator Frank Carlson in 1967, and in 1969 Roberts became administrative assistant to Kansas Congressman Keith Sebelius.

After Keith Sebelius announced his retirement Roberts easily won the Republican primary, which was tantamount to election in the heavily Republican 1st District. He was reelected seven times without serious difficulty, never received less than sixty percent of the vote, and in 1988 he ran unopposed.

After the retirement of Senator Nancy Kassebaum Roberts easily won the Republican primary, and in the general election he defeated Democratic State Treasurer Sally Thompson - almost certainly helped by the presence of Bob Dole atop the ticket as the Republican presidential candidate. No Democratic candidate opposed Roberts in 2002, allowing him to win re-election to a second term, and he went on to win a third term in 2008.

Although Roberts is the dean of the Kansas Congressional Delegation he is actually the state's junior Senator, since Sam Brownback was sworn in on the night of the election in 1996 for the balance of Dole's Senate term. Roberts is to become Kansas' senior Senator in the 112th Congress, as Brownback is retiring from the Senate to run for Governor.

Roberts' voting record is decidedly conservative. Among other issues he is pro-life, opposes same-sex marriage, supports the Patriot Act, and is for loosening restrictions on NSA wiretapping.

As chairman of the Senate Select Committee on Intelligence, Roberts was responsible for the committee's investigation into the intelligence failures prior to the 2003 invasion of Iraq. The first half of the Senate Report of Pre-war Intelligence on Iraq was released on July 9, 2004 and the second half, according to language voted on by the full Committee, consists of five parts including whether public statements and reports and testimony regarding Iraq by U.S. Government officials made between the Gulf War period and the commencement of Operation Iraqi Freedom were substantiated by intelligence information, postwar findings about Iraq's weapons of mass destruction programs and links to terrorism and how they compare with prewar assessments, prewar intelligence assessments about postwar Iraq, intelligence activities relating to Iraq conducted by the

Policy Counterterrorism Evaluation Group (PCTEG) and the Office of Special Plans within the Office of the Under Secretary of Defense for Policy, and the use by the Intelligence Community of information provided by the Iraqi National Congress (INC).

Almost two years after finishing Phase I of the investigation Roberts released the Committee's schedule for completion of Phase II, and in March of 2006 he said, "Today members of the Committee were provided three draft reports of the Phase II inquiry including postwar findings about Iraq's weapons of mass destruction programs and links to terrorism and how they compare with prewar assessments, the use by the Intelligence Community of information provided by the Iraqi National Congress, and prewar intelligence assessments about postwar Iraq. Then in August of 2006 Roberts publicly released the findings of fact and conclusions of the first two reports.

Roberts was one of nine Senators to vote against the Detainee Treatment Act of 2005, and in September of 2006 he voted with a largely Republican majority to suspend habeas corpus provisions for anyone deemed by the Executive Branch an "unlawful combatant," and barring them from challenging their detentions in court. The vote authorized the President to establish permissible interrogation techniques and to "interpret the meaning and application" of international Geneva Convention standards, so long as the coercion fell short of 'serious' bodily or psychological injury." The bill became law on October 17, 2006.

JIM SASSER
Senator from Tennessee

James Ralph Sasser is a Democratic politician and attorney who served three terms as a United States Senator from Tennessee from 1977 to 1995. He was Chairman of the Senate Budget Committee, and from 1995 to 1999, during the Clinton Administration, he was the United States Ambassador to the People's Republic of China.

Sasser was born in Memphis, Tennessee on September 30, 1936, attended public schools in Nashville, and studied at the University of Tennessee from 1954 to 1955. He earned his undergraduate degree from Vanderbilt University in 1958, followed by his law degree from Vanderbilt Law School in 1961, and was admitted to the Tennessee Bar in 1961. He then began practicing law in Nashville, and from 1957 to 1963 served in the Marine Corps Reserve.

A longtime Democratic activist, Sasser was manager of Albert Gore, Sr.'s unsuccessful 1970 reelection campaign. He then sought election in his own right, won his party's 1976 nomination for the Senate, and set out to attack the record of one-term incumbent Senator Bill Brock, who was heir to a Chattanooga candy fortune. Sasser emphasized Brock's connections to former President Richard M. Nixon and his use of income tax code provisions which had, despite his great wealth and considerable income, resulted in his paying less than two thousand dollars in income tax the previous year. Sasser was able to capitalize on the tax issue by pointing out Brock had paid less than many Tennesseans of considerably more modest means.

Sasser's campaign was also greatly aided by the efforts of ex-Senator Gore. Brock had defeated the elder Gore for the Senate in 1970 largely upon the basis of Gore's support for civil rights, his friendship with the Kennedy political family, and his opposition to the Vietnam War. In the end Sasser won rather handily over Brock, and went on to serve three Senate terms.

With the retirement of Senator Lawton Chiles in 1989, Sasser became Chairman of the Senate Budget Committee. In that role he served as a key ally of Senate Majority Leader George Mitchell of Maine and helped negotiate the 1990 budget summit agreement with President George H. W. Bush. With this success under his belt he began to work his way upward in the party leadership, and when Leader Mitchell announced his intention to retire it seemed to be a foregone conclusion that upon his re-election in 1994 Sasser would become the new majority leader.

There were two unforeseen events which negated that destiny. One was the large scale of discontent the American people seemed to have toward the first two years of the Clinton administration, especially the proposal for a national health-care system largely put together and advocated by Clinton's wife, Hillary Rodham Clinton. The other was the somewhat unexpected nomination of Nashville heart transplant surgeon William Frist by the Republicans.

Frist was a political unknown and a total novice at campaigning, but because he was from one of Nashville's most prominent and wealthiest medical families he had name recognition in the Nashville area and resources to match the campaign war chest built up by a three-term incumbent. Another factor working to Frist's advantage was the simultaneous Republican campaign by actor and attorney Fred Thompson for the other Tennessee Senate seat, which had come open when Al Gore Jr. resigned to become Vice President. To an extent Frist was able to bask in the reflected glory of Thompson's formidable stage presence, and eventually developed campaigning skills which were almost totally absent in the early stages of the contest. Another factor in Frist's favor was Sasser was never seen as possessing much charisma of his own, and during the campaign Nashville radio stations were derisive towards him to the point of saying he could only win "a Kermit the Frog lookalike contest." In the November 1994 general election, in one of the largest upsets on a night filled with them, Frist defeated Sasser quite decisively.

Sasser went on to serve as Ambassador to China during the period of alleged nuclear spying and the campaign finance controversy that involved possible efforts by China to influence domestic U.S. politics during the Clinton Administration. Sasser again gained media attention when the U.S. Embassy in Beijing was besieged after U.S. warplanes mistakenly bombed the Chinese Embassy in Belgrade during the U.S. intervention in the Kosovo War. Shortly after the siege of the Embassy was lifted, Ambassador Sasser retired (he was slated to do so before the siege, so his retirement was not a direct result) and returned to the United States, where he presently divides his time between Tennessee and Washington, D.C. as a consultant.

FRANK M. TEJEDA
Congressman from Texas

Frank Mariano Tejeda (Oct 2, 1945 - Jan 30, 1997) was a decorated Marine and Democratic politician from Texas who served in the Texas House of Representatives, the Texas Senate, and in the United States House of Representatives.

Tejeda was born in San Antonio, Texas, attended St. Leo's Catholic School, and later graduated from Harlandale High School. He served in the Marine Corps from 1963 to 1967, and was wounded in action during the Vietnam War. Tejeda was decorated with the Silver Star, Bronze Star, and Purple Heart, and attained the rank of Major in the Marine Corps Reserve.

After his Marine Corps service Tejeda completed his Bachelor's degree at St. Mary's University in San Antonio, and earned a J.D. from University of California, Berkeley Law School in 1974.

Tejeda began his political career in the Texas Legislature, serving in the Texas House from 1976 to 1987 and then the Texas Senate from 1987 to 1993. While serving in the legislature he earned two Masters Degrees - in 1980 he received an M.A. from Harvard University, and in 1989 an LL.M. from Yale University's Law School.

Tejeda was elected with eighty-seven percent of the vote to the U.S. Congress in 1992, and represented the 28th Congressional District of Texas. While in the House he served on the Armed Services and the Veterans' Affairs

Committees, and his work in Congress focused on veterans' issues.

In January of 1997, shortly after the beginning of his third term, Congressman Tejeda died after a year-long battle with brain cancer and was buried with full military honors at Fort Sam Houston National Cemetery in San Antonio.

The VA outpatient clinic in San Antonio, Texas was posthumously named in Tejeda's honor, and after his death the Marine Corps Reserve Association created the "Major Frank M. Tejeda Leadership Award" to recognize leaders committed to the Marine Corps.

The NEISD Middle School in San Antonio was also posthumously named in his honor, and in 1997 U.S. Highway 281 from Interstate 410 to the Atascosa/Bexar County line was re-named the "Congressman Frank M. Tejeda Memorial Highway" by the Texas Legislature.

CRAIG THOMAS
Senator from Wyoming

Craig Lyle Thomas (Feb 17, 1933 - June 4, 2007) was a politician who served for over twelve years as a Republican United States Senator from Wyoming. In the Senate Thomas was considered an expert on agriculture and rural development, and he served in key positions in several state agencies, including a long tenure as Vice President of the Wyoming Farm Bureau from 1965 to 1974. Thomas resided in Casper for twenty-eight years, and in 1984 was elected from Casper to the Wyoming House of Representatives, where he served until 1989.

When in 1989 Dick Cheney, who occupied Wyoming's only seat in the House of Representatives, resigned to become Secretary of Defense, Thomas became the Republican candidate to succeed him and won the ensuing special election. He was re-elected in 1990 and 1992, and in 1994 he ran for and won the Senate seat being vacated by fellow conservative Republican Malcolm Wallop by defeating popular Democratic Governor Mike Sullivan. He was re-elected in 2000 and 2006, easily beating Democratic candidates in both elections with margins of seventy percent or more.

Thomas was born and reared in Cody, the seat of Park County in northwestern Wyoming, roughly fifty miles east of Yellowstone National Park. His parents were public school teachers who operated a dude ranch business on the edge of

Yellowstone during the summers, and the family's interest in tourism later led Thomas to purchase a small hotel in Torrington.

Thomas graduated from the University of Wyoming in Laramie with a degree in animal husbandry, and thereafter served as an officer in the Marine Corps from 1955 to 1959 where he attained the rank of Captain. He also obtained a law degree from La Salle Extension University, although he did not list it on later official biographies.

As chairman of the National Parks Subcommittee, Thomas authored legislation to provide funding and management reforms to protect America's national parks into the 21st century. For this and other relevant legislation, he was honored by the National Parks and Conservation Association with their William Penn Mott, Jr., Park Leadership Award, as well as the National Parks Achievement Award. As the senior member of the Senate's influential Finance Committee, Thomas had been involved in issues such as Social Security, trade, rural health care and tax reform, and as co-chair of the Senate Rural Health Caucus Senator Thomas worked on legislation to improve health care opportunities for rural families.

Thomas entered the hospital shortly before ballots were to be cast in November of 2006 and was initially treated for pneumonia, but two days after the election a diagnosis of leukemia was announced. He immediately underwent treatment in the form of chemotherapy, and then returned to work in December - a month earlier than expected. In early 2007 Thomas said he was feeling better than he had in a long time, but returned to the hospital for a second round of chemotherapy a month later. In mid-June he was reported to be in serious condition, and was struggling with an infection while undergoing a second round of chemotherapy at

Bethesda Naval Medical Center in Maryland. He was pronounced dead that same day from complications of leukemia.

Thomas' services were held in the Methodist Church in Casper on June 9, 2007 and the two Senate leaders, Majority Leader Harry Reid (D-NV) and Minority Leader Mitch McConnell (R-KY), headed a delegation of some twenty members of Congress who came to pay their respects.

Under Wyoming law Governor Dave Freudenthal was required to appoint a new Senator from a list of three submitted by the Wyoming Republican Party's central committee because the seat was vacated by a Republican. The GOP met in Casper to select three candidates from thirty applicants and nominated Tom Sansonetti, former State Treasurer Cynthia Lummis, and State Senator John Barrasso. Three days later Governor Freudenthal appointed Barrasso as Thomas' successor in the U.S. Senate.

Senator Craig Thomas has been honored posthumously by having the Visitor Center in Grand Teton National Park named for him. The building, which is in Moose, Wyoming, was dedicated in August of 2007 with many dignitaries attending, including Vice President Dick Cheney.

WILLIAM M. TUCK
Governor of and Congressman from Virginia

William Munford Tuck (Sept 28, 1896 - June 9, 1983) was a Democrat who served as the 55th Governor of Virginia from 1946 to 1950.

Tuck was the youngest son of Halifax County Virginia tobacco warehouseman Robert James Tuck and Virginia Susan Fritts. He graduated from the College of William and Mary, earning a teacher's certificate, and served in the Marine Corps in the Caribbean in 1917. He graduated from Washington and Lee University Law School in 1921, and was then admitted to the Virginia bar.

Tuck was elected as a Democrat to a U.S. Congress seat in 1953 to fill the vacancy created by Thomas Bahnson Stanley, who had resigned to run for Governor. There he opposed most major items of civil rights legislation during the 1950s and 1960s, and also promised "massive resistance" to the Supreme Court's 1954 decision banning segregation, *Brown v. Board of Education*. Tuck was a delegate to Democratic National Conventions of 1948 and 1952, and also served in both houses of the Virginia General Assembly and as Lieutenant Governor of Virginia from 1942 to 1946. As Governor he reorganized state government, enacted a right-to-work law, and created a state water pollution control agency.

William Tuck is buried in Oak Ridge Cemetery in South Boston, Virginia.

177

JOHN WARNER
Senator from Virginia
Secretary of the Navy

John William Warner KBE is a Republican politician who served as Secretary of the Navy from 1972 to 1974 and as a five-term United States Senator from Virginia from 1979 to 2009. Warner was once married to actress Elizabeth Taylor, and is a veteran of World War II and the Korean War.

Warner was born on February 18, 1927 to John W. and Martha Budd Warner and grew up in Washington, D.C., where he attended the elite St. Albans School, but he graduated from Woodrow Wilson High School (a D.C. Public School) in February of 1945.

He enlisted in the United States Navy in January of 1945, shortly before his eighteenth birthday, served until the following year, and left as a Petty Officer 3rd Class. He then went to college at Washington and Lee University, graduated in 1949, and then entered the University of Virginia Law School.

Warner joined the Marine Corps in October of 1950 after the outbreak of the Korean War and served in Korea as a ground officer with the 1st Marine Aircraft Wing. He continued in the Marine Corps Reserves after the war, and reached the rank of Captain. He then resumed his studies, taking courses at George Washington University, and received his law degree in 1953. He then became a law clerk to Chief Judge E. Barrett Prettyman of the United States Court of Appeals, in 1956 became an assistant U.S. attorney, and in 1960 entered private law practice.

In February of 1969 Warner was appointed Undersecretary of the Navy in the Nixon administration, and in 1972 he succeeded fellow Marine John H. Chafee as Secretary of the Navy. In that capacity he participated in the Law of the Sea talks, and negotiated the Incidents at Sea Executive Agreement with the Soviet Union. He was subsequently appointed by Gerald Ford to the post of Director of the American Revolution Bicentennial Administration.

Warner entered politics in the 1978 Virginia election for the U.S. Senate. Known primarily as Elizabeth Taylor's husband, he finished second at the state Republican convention to Richard Obenshain, but when Obenshain died in a plane crash two months later Warner was chosen to replace him and narrowly won the general election over Democrat Andrew P. Miller, the former Attorney General of Virginia. He remained in the Senate until January 3, 2009, was the second-longest serving Senator in Virginia's history behind only Harry F. Byrd, Sr., and was by far the longest-serving Republican. In August of 2007 Warner announced he would not seek re-election in 2008.

His committee memberships included the Environment and Public Works Committee, the Senate Committee on Health, Education, Labor, and Pensions, and the Senate Select Committee on Intelligence. Most importantly, as chairman of the Senate Armed Services Committee, he protected and enlarged the flow of billions of dollars into the Virginia economy each year via the state's Naval installations and shipbuilding firms.

Warner was considerably more moderate than most Republican Senators from the South. He was among the minority of Republicans to support gun control laws, voted for the Brady Bill, and in 1999 was one of only five Republicans to vote to close the so-called "gun show loophole." Then in 2004 Warner was one of three Republicans to sponsor an amendment by Senator Dianne Feinstein (D-CA) which sought a ten-year extension of the Assault Weapons Ban.

Warner was considered "pro-choice" and supports embryonic stem cell research, although he received high ratings from pro-life groups because he voted in favor of many abortion restrictions. In 2004 he was among the minority of his party to vote to expand hate crime laws to include sexual orientation as a protected category, and while he supports a Constitutional Amendment banning same-sex marriage, he raised concerns about the most recent Federal Marriage Amendment as being too restrictive as it would have potentially banned civil unions as well.

In 1987 Warner was one of the few Republicans who crossed party lines to reject the nomination of fellow Marine Robert Bork to the Supreme Court by President Ronald Reagan.

In 1994 Warner campaigned for former state Republican Attorney General turned Independent candidate Marshall Coleman against fellow Republican and Marine Oliver North in North's unsuccessful campaign to unseat Virginia's Democratic Senator, Chuck Robb – yet another Marine. North's loss to Robb was very close, with Coleman finishing in single digits and looking like a spoiler, and Warner's actions were seen as the direct cause of a fellow Republican's defeat.

Because of his centrist stances on many issues and his snubbing of fellow Republicans, Warner faced opposition from angry members of his own party when he decided to run for a fourth term in the Senate in 1996. Many of Virginia's staunch Republican voters began a "Dump Warner" campaign to try to deny him re-nomination, but Virginia's GOP party rules allow the incumbent to select the nominating process, and since Warner knew he would probably lose at a convention or caucus where only party regulars would be voting, he selected a primary. In Virginia primaries are open to all registered voters, so he encouraged Democrats and independents to vote for him in the primary. The strategy worked, and Warner handily defeated Republican rival James C. Miller III for the nomination.

In the general election that year Warner was expected to win in a cakewalk over relatively unknown (at that time) Democrat Mark Warner (no relation), who had never held elective office. The election turned out to be much closer than many pundits had expected, and the Democrat was able to tighten the race by taking full advantage of the discontent of conservative Republican voters. Even though he lost, the close election provided Mark Warner enough momentum to successfully run for Governor of Virginia five years later.

According to George Stephanopoulos, a former close aide to President Bill Clinton, Warner was among the top candidates to replace Les Aspin as Secretary of Defense in the Clinton administration before Clinton selected William Perry. During Clinton's second term William Cohen of Maine, another moderate Republican Senator, held this position.

Warner was among ten GOP Senators who voted against the charge of perjury during Clinton's impeachment (the others were Richard Shelby of Alabama, Ted Stevens of

Alaska, Susan Collins of Maine, Olympia Snowe of Maine, John Chafee of Rhode Island, Arlen Specter of Pennsylvania, Jim Jeffords of Vermont, Slade Gorton of Washington and Fred Thompson of Tennessee). Warner and others who voted against the article angered many Republicans by their position, but unlike Snowe, Collins, Specter, Jeffords and Chafee, he and the rest of the Republicans voted "guilty" on the second article.

In October of 2007 Warner was admitted to Inova Fairfax Hospital and underwent surgery to correct atrial fibrillation, or an irregular heartbeat. He made a full recovery and returned to work the following week, although in February of 2008 he was again admitted for a scheduled observation of his heart condition.

On December 12, 2008 the Office of the Director of National Intelligence awarded Warner the first ever National Intelligence Distinguished Public Service Medal, and on January 8, 2009 the Secretary of the Navy announced it would name the next Virginia-class submarine after John Warner. *USS John Warner* (SSN-785) will be the twelfth Virginia-class submarine. Then on February 19, 2009 the British Embassy in Washington, D.C., announced Queen Elizabeth II would name Warner an honorary Knight Commander for his work in strengthening the American-British military alliance. As a non-British citizen, the title of Knight Commander of the Most Excellent Order of the British Empire allows Warner to put the Post-nominal letters KBE after his name.

JAMES H. "JIM" WEBB

Senator from Virginia
Secretary of the Navy

James Henry "Jim" Webb, Jr. is the senior Senator from Virginia. He is also an author and a former Secretary of the Navy under President Ronald Reagan.

A 1968 graduate of the U.S. Naval Academy, Webb served as a Marine Corps infantry officer until 1972 and is a highly decorated Vietnam War combat veteran. During his four years with the Reagan administration, Webb served first as the Assistant Secretary of Defense for Reserve Affairs, then as Secretary of the Navy.

Webb won the Democratic nomination for the 2006 Virginia Senate race by defeating Harris Miller in the primary, and then won the general election by defeating Republican incumbent George Allen. Webb's thin margin in the general election (less than 0.5%) kept the outcome uncertain for nearly two days after polls closed on November 7, 2006, and provided the final seat that tilted the Senate to Democratic control.

Webb was born on February 9th, 1946 in Saint Joseph, Missouri, to James Henry Webb and his wife Vera Lorraine Hodges. He grew up in a military family, descended from Scots Irish immigrants from Ulster who emigrated in the 18th century to the British North American colonies. Webb's 2004 book *Born Fighting: How the Scots-Irish Shaped*

America details his family history, noting that his ancestors fought in every major American war.

Webb's father, a career officer in the U.S. Air Force, flew B-17s and B-29s during World War II, dropped cargo during the Berlin Airlift, and was later involved in missile programs. Because of his father's military career Webb grew up on the move, attending more than a dozen schools across the U.S. and in England. After graduating from high school in Bellevue, Nebraska he attended the University of Southern California on a Navy Reserve Officer Training Corps scholarship from 1963 to 1964, and in 1964 earned appointment to the United States Naval Academy in Annapolis. At Annapolis Webb was a member of the Brigade Honor Committee. He also won a varsity letter for boxing, at one point fighting a controversial match against Oliver North, which was won by North on decision. He graduated in 1968 in the same class with North, Dennis C. Blair, Michael Mullen, and future Commandant of the Marine Corps Michael Hagee.

After graduating from the Naval Academy Webb was commissioned a Second Lieutenant in the Marine Corps, and as a First Lieutenant during the Vietnam War he served as a platoon commander with Delta Company, 1st Battalion 5th Marines. He earned a Navy Cross, the second highest decoration in the Navy and Marine Corps, for heroism in Vietnam, as well as the Silver Star, two Bronze Stars, and two Purple Hearts.

In a November 19, 2006 appearance on *Meet the Press,* Webb told host Tim Russert, "And I, you know, I'm one of these people who - there, there aren't many of us - who can still justify for you the reasons that we went into Vietnam, however screwed up the strategy got."

Webb attended Georgetown Law Center from 1972 to 1975, graduating with a Juris Doctor degree, and while there wrote his first book, *Micronesia and U.S. Pacific Strategy.* From 1977 to 1981 he worked on the staff of the House Committee on Veterans Affairs, and during this time he also represented veterans pro-bono and taught at the Naval Academy.

During the Reagan Administration Webb served as the nation's first Assistant Secretary of Defense for Reserve Affairs from 1984 to 1987, and during that time he sought to reorganize the Marine Corps. He was gravely concerned with the disarray the Marines had fallen into post-Vietnam. Drug use, racial infighting, and low morale within the Corps left him with the impression it was no longer America's premier fighting force. The Marine Corps was also rocked by two scandals during this time - the Clayton Lonetree espionage affair, where Lonetree became the first Marine convicted of espionage, and Lieutenant Colonel Oliver North's central role in the Iran-Contra affair.

In 1987 he was named Secretary of the Navy and became the first Academy graduate to serve as the civilian head of the Navy. As Secretary, Webb pushed the appointment of Alfred M. Gray, Jr. as Commandant of the Marine Corps, hoping that Gray could reshape the Corps into the elite unit it once was. Webb then resigned in 1988 after refusing to agree to reduce the size of the service. He had wished to increase the Navy to 600 ships, and continued his opposition to the expanded role for women in the Navy after his appointment. In a speech at the Naval Academy he referred to female students as "thunder thighs," much to the delight of those who opposed increased opportunity for women (who became known as the "Webbites"). As revealed in *The Reagan*

Diaries, President Ronald Reagan wrote on February 22, 1988, "I don't think the Navy was sorry to see him go."

After his resignation Webb earned his living primarily as an author and filmmaker, and he won an Emmy Award for his 1983 PBS coverage of the U.S. Marines in Beirut.

During the 2004 presidential campaign Webb wrote an op-ed piece for *USA Today* in which he, as a military veteran, evaluated the candidacies of John Kerry and George W. Bush. He criticized Kerry for the nature of his opposition to the Vietnam War in the 1970s while affiliated with the Vietnam Veterans Against the War, and accused Bush of using his father's connections to avoid service in Vietnam. Webb also wrote that Bush had "committed the greatest strategic blunder in modern memory" with the 2003 invasion of Iraq.

In 1994 Webb endorsed incumbent Democrat Charles Robb, a former Marine, for reelection to the Senate over Webb's former Naval Academy classmate and fellow Marine Oliver North. Then in 2000 he endorsed Republican George Allen over Robb, and in 2006 he ran against Allen himself after a late 2005 "draft Webb" campaign began on the Internet.

Webb benefited from the fallout from an incident in which Allen used the word "macaca" to refer to S.R. Sidarth, who was filming an event as a "tracker" for the Webb campaign. A poll the following week showed Webb gaining ten percentage points and the race, which at one point looked like a sure win for Allen, became one of the most watched and closest contests of the 2006 elections.

Allen had been expected to be reelected relatively easily, and it was thought reelection would prepare him for a possible 2008 Presidential candidacy, but Webb's entry into the race changed the political landscape. Political analyst

The Marines Have Landed

Larry Sabato said in May of 2006 that "Jim Webb is George Allen's worst nightmare: a war hero and a Reagan appointee who holds moderate positions... Allen tries to project a Reagan aura, but Webb already has it." In September *Bloomberg's* Catherine Dodge wrote an article highlighting Webb and the Senate race, and said "Webb isn't a typical Democrat. His family hails from the rural southern part of the state. He's pro-gun ownership, and takes a harder line on illegal immigration than many Senate Republicans."

Five female graduates of the Naval Academy held a press conference, decrying a 1979 article by Webb titled *Women Can't Fight*. The women said Webb's article contributed to an atmosphere of hostility and harassment towards women at the Academy, although Webb was later endorsed by nine military women who stated he was a "man of integrity who recognizes the crucial role that women have in the military today."

On November 9th, after the AP and Reuters projected Webb had won the seat, Allen conceded the election, and on November 15th Senate majority leader in waiting Harry Reid assigned Webb to three committees: the committees on Foreign Relations, Veterans' Affairs, and Armed Services. That same day, an op-ed authored by Webb appeared in the pages of the *Wall Street Journal*. Titled *Class Struggle*, the piece addressed what Webb believes is a growing economic inequality in the United States, touching on what he feels are overly permissive immigration policies, extravagant executive compensation, the detrimental effects of free trade and globalization, iniquitous tax cuts, and speedily rising health care costs. He then went on to attack the "elites" who he says perpetuate the aforementioned woes for their personal economic gain.

On November 28, 2006, at a White House reception for those newly elected to Congress, Webb declined to stand in the line to have his picture taken with the President, whom he had often criticized during the campaign. According to Congressman Jim Moran of Virginia, aides had warned the President to be "extra sensitive about talking to Webb about his son, since Webb's son had had a recent brush with death in Iraq." The President approached Webb later and asked him, "How's your boy?," referring to Webb's son, a Marine serving in Iraq. Webb replied, "I'd like to get them out of Iraq, Mr. President." Bush responded, "That's not what I asked you. How's your boy?" Webb responded, "That's between me and my boy, Mr. President." *The Hill* cited an anonymous source who claimed Webb was so angered by the exchange that he confessed he was tempted to "slug" the President. Webb later remarked in an interview, "I'm not particularly interested in having a picture of me and George W. Bush on my wall."

In response to the incident some conservatives criticized Webb, including George Will, who called Webb a "boor" and wrote, "(Webb) already has become what Washington did not need another of, a subtraction from the city's civility and clear speaking." Others, such as conservative columnist and former Reagan speechwriter Peggy Noonan, reserved their criticism for Bush, writing, "I thought it had the sound of the rattling little aggressions of our day, but not on Mr. Webb's side."

On January 4, 2007 Webb was sworn into the 110th U.S. Senate, accompanied by Senator John Warner (R-VA), a fellow former Marine and Secretary of the Navy, and former Virginia Democratic Senator Charles Robb, a Marine as well. Later that same day Webb was asked about the exchange with President Bush in an appearance on *Hardball with Chris Matthews.* He told Matthews, "My feeling about that - first of all, it's been kind of a bit overblown. But I think when people are now seeing how John McCain is handling the situation with his son being in the Marine Corps, perhaps they can understand a little bit more what I was having to go through during the entire campaign. I greatly respect my son's service and all of the people who are serving. At the same time, I have not commented, even to many of my friends, about the operational side. That's personal to me in terms of my feelings about it. And it was not a casual comment. As I said in the piece that you just ran, I think the best article that was written on that was by Peggy Noonan in the *Wall Street Journal* when she basically said that the lack of civility was not mine and I feel that way." After his son returned from Iraq Webb "buried the hatchet" with the President by setting up a private chat with his son, the President and himself in the Oval Office.

Webb's first legislative act was to introduce a bill, the Post-9/11 Veterans Educational Assistance Act, to expand benefits for military families. The act replaces key provisions of the Montgomery G.I. Bill for recent veterans and "makes veterans benefits identical to those soldiers received following World War II." He said, "With many of our military members serving two or three tours of duty in Iraq and Afghanistan, it is past time to enact a new veterans' education program modeled on the World War II era G.I. Bill. This is exactly what our legislation does." It became law on June 30, 2008.

On January 23, 2007 Webb delivered the Democratic response to the President's State of the Union address, focusing on the economy and Iraq. Webb's speech drew positive reviews, and was regarded as one of the stronger State of the Union responses in recent memory. As a decorated war veteran, he spoke of his family's military past, his own passionate attachment to the military, and the way in which previous Presidents had always attempted to ensure all precautions had been taken when sending young Americans into harm's way.

On March 26, 2007 a Senatorial aide of Webb, Phillip Thompson, was arrested for carrying Webb's loaded pistol as he entered the Russell Senate Office Building and for carrying unregistered ammunition. The weapon was discovered when Thompson went through an X-ray machine with a briefcase which contained a loaded pistol and two additional loaded magazines. Charges against the aide were dismissed after prosecutors concluded it could not be proven beyond a reasonable doubt that Thompson was aware the gun and ammunition were in the briefcase. Webb responded to his aide's arrest by reiterating his support for gun-owners' rights, saying, "I'm a strong supporter of the Second Amendment. I have had a permit to carry a weapon in Virginia for a long time, and I believe that it's important. It's important to me personally, and to a lot of people in the situation that I'm in, to be able to defend myself and my family."

In August Webb paid a visit to Vietnam as part of a two-week trip to five Southeast Asian countries. The Senator from Virginia, who serves as chair of the Senate Foreign Relations subcommittee on East Asia and Pacific Affairs, stopped in Hanoi, Da Nang, and Ho Chi Minh City on August 19, where he met government officials, business leaders, and friends from his decades of close involvement in U.S.-Vietnam relations. Senator Webb, who can speak Vietnamese, has worked and traveled throughout this vast region from Micronesia to Burma for nearly four decades as a Marine Corps officer, defense planner, journalist, novelist,

Department of Defense executive, and as a business consultant, and in the 1990's he worked as a consultant for companies wishing to do business in Vietnam.

In January of 2010 Webb condemned Vietnam's jailing of four dissidents, but urged the Obama administration not to isolate the communist nation. Webb voiced concern about Vietnam's jailing of the four for subversion in a day-long trial the previous week. He stated, "The arrest and trial of these individuals illustrates the growing pressure in Asia towards government censorship and authoritarian control. Rather than isolate Vietnam for its actions, I encourage the Obama administration to continue to raise the issues of freedom of association and the rule of law with the government of Vietnam."

Webb was frequently mentioned as a possible Vice Presidential Democratic nominee for Barack Obama in 2008 due to his military experience and moderate policy positions. Although he said he was not interested in the Vice Presidency, speculations about him being picked by Obama, the presumptive Democratic nominee at the time, were still heard.

Some felt that his commentary on women serving in the military (e.g., his article *Women Can't Fight*) was a strong consideration as to his possible candidacy. His selection would have closely followed the somewhat divisive Democratic primary battle between Obama and Hillary Clinton, whose candidacy had received strong support from organized feminism, and who would have been the first female major party nominee for the presidency had she won. This situation may have made the prospect of Webb as Obama's running mate politically untenable, as it could have caused many Democratic Clinton supporters to balk at switching allegiance to Obama.

Webb is also an author of many books, stating, "I've written for a living all my life, so writing is as much a part of me as working out." His successful first novel, 1978's *Fields of Fire*, which was drawn from personal experience, tells the story of a platoon of Marines in late 1960s Vietnam. Reviewers hailed its pull-no-punches descriptions of infantry life and combat. After five more novels, including *A Sense of Honor, A Country Such as This, Something to Die For, The Emperor's General,* and *Lost Soldiers*, he wrote a work of nonfiction, *Born Fighting: How the Scots-Irish Shaped America,* tracing the role people of Scots-Irish ancestry have played in American history and culture. Webb argues that, contrary to the "cracker" and "redneck" stereotypes often applied to the Scots-Irish, many of whom settled in Appalachia, the American Midwest and the American South, the Scots-Irish were central to defining American working class values and culture. He lauds their fierce independent streak and individualism, and explains how their political pragmatism has often led them to play the role of swing voters in elections - for example as Reagan Democrats, and as voters for Ross Perot and the Reform Party.

Webb also wrote the story for and was the executive producer of the 2000 movie *Rules of Engagement*, which starred Tommy Lee Jones and Samuel L. Jackson.

Among Webb's awards for community service and professional excellence are the Department of Defense Distinguished Public Service Medal, the Medal of Honor Society's Patriot Award, the American Legion National Commander's Public Service Award, the Veterans of Foreign Wars Media Service Award, the Marine Corps League's Military Order of the Iron Mike Award, the John H. Russell Leadership Award, and the Robert L. Denig Distinguished Service Award.

PETE WILSON
Governor of and Senator from California

Peter Barton "Pete" Wilson is a Republican who served as the 36th Governor of California, the culmination of more than three decades in the public arena which included eight years as a United States Senator, eleven years as Mayor of San Diego and five years as a California State Assemblyman.

Wilson was born on August 23, 1933 in Lake Forest, Illinois, a suburb north of Chicago, to James Boone Wilson and Margaret Callaghan Wilson. His father was originally a jewelry salesman who later became a successful advertising executive. The Wilson family moved to St. Louis, Missouri when Pete was in junior high school, and there he attended the St. Louis Country Day School, an exclusive private high school, where he won an award in his senior year for combined scholarship, athletics, and citizenship. In the fall of 1952 Wilson enrolled at Yale University in Connecticut, where he received a U.S. Navy (Marine Corps) ROTC scholarship, majored in English, and earned his Bachelor of Arts degree.

After graduation from Yale Wilson served for three years in the Marine Corps as an infantry officer and eventually became a platoon leader. Upon completion of his Marine Corps service, Wilson earned a law degree from the Boalt Hall School of Law at the University of California, Berkeley.

193

In 1962, while working for Republican gubernatorial candidate Richard M. Nixon, Wilson got to know one of Nixon's top aides, Herb Klein. Klein suggested Wilson might do well in Southern California politics, so in 1963 he moved to San Diego.

Wilson began his practice as a criminal defense attorney in San Diego, but soon found such work to be low-paying and personally repugnant. He later commented to the *Los Angeles Times*, "I realized I couldn't be a criminal defense lawyer, because most of the people who do come to you are guilty." Wilson switched to a more conventional law practice and continued his involvement in local politics, working for Barry Goldwater's unsuccessful Presidential campaign in 1964. His like for politics and managing the day-to-day details of the political process was growing. He put in long hours for the Goldwater campaign, earning the friendship of local Republican boosters, and in 1966, at the age of thirty-three, he ran for and won a seat in the California State Legislature.

As the Mayor of San Diego Wilson guided the city as it transformed from a quiet U.S. Navy and Marine Corps town to an international trade hub, and is credited with amending the city charter to make public safety the first and foremost responsibility of city government and leading an effort to manage San Diego's dynamic growth and revitalize the city's downtown area. He substantially cut the property tax rate, and imposed a limit on the growth of the city budget which became a model for California's subsequently adopted Proposition 13. Wilson was largely responsible for beginning the downtown transformation of the Gaslamp Quarter from a drug-infested area to a highly business friendly and successful downtown, and coined the slogan for San Diego which is still widely used today – "San Diego: America's finest city."

In 1982 Wilson won the Republican primary in California to replace retiring U.S. Senator S. I. Hayakawa, and his Democratic opponent was outgoing two-term Governor Jerry Brown. Wilson was known as a fiscal conservative who supported Proposition 13 while Brown opposed it, however Brown ran on his record as the Governor who had built the largest state budget surpluses in California history. Both Wilson and Brown were moderate-to-liberal on social issues, including support for abortion rights, and the election was expected to be close, with Brown holding a slim lead in most polls leading up to Election Day. Wilson hammered away at Brown's appointment of the (liberal) California Chief Justice Rose Bird, using this to portray himself as tougher on crime. Brown's late entry into the 1980 Democratic Presidential primary, after promising he would not run, was also an issue. Finally President Ronald Reagan made a number of visits to California late in the race to campaign for Wilson. Despite exit polls indicating a narrow Brown victory, in the end Wilson edged him out to win the election.

Reagan was repaid for his support and the new Senator was able to demonstrate his Marine Corps mettle when, in 1985, Wilson cast a key vote in favor of the Reagan budget. Just thirty-two hours after having surgery to remove his ruptured appendix, Wilson arrived by ambulance at the Capitol Building shortly after midnight and was wheeled onto the Senate floor wearing blue pajamas covered by a brown velour robe. Not only was Wilson able to cast his vote in a firm voice, he even held a brief press conference during the late-night session in which he jokingly asked reporters, "What are you all doing up this late?"

As a United States Senator from 1983 to 1991 Wilson was a vocal proponent for a stronger, more military-based defense and U.S. foreign policy, and as a member of the Senate Armed Services Committee he called for early implementation of President Reagan's Strategic Defense Initiative, the national ballistic missile defense system popularly known as "Star Wars." Wilson also cosponsored the Federal Intergovernmental Regulatory Relief Act requiring the Federal government to reimburse states for the cost of new Federal mandates. A fiscal conservative, he was named the Senate's "Watchdog of the Treasury" for each of his eight years in the nation's capitol.

Wilson was re-elected to a second term in the U.S. Senate in 1988, but decided to run for Governor of California during the first two years of his term and resigned from the Senate in 1990 when he won the election. He defeated former San Francisco Mayor Dianne Feinstein, who would go on to be elected to Wilson's former U.S. Senate seat two years later.

Wilson's eight years as the Governor saw California go into a strong economic recovery. Inheriting the state's worst economy since the Great Depression, Wilson insisted on strict budget discipline and worked to rehabilitate the state's environment for investment and new job creation. His term saw market-based, unsubsidized health coverage made available for employees of small businesses and additional anti-fraud measures implemented which were credited with reducing workers' compensation premiums by as much as forty percent.

Wilson led efforts to enact tougher crime measures and signed into law the highly popular "Three Strikes" (25 years to life for repeat felons) and "One Strike" (25 years to life upon the first conviction of aggravated rape or child molestation) statutes. Wilson also supported resuming the death penalty in California, and after a twenty-five year moratorium signed a death warrant for the execution of Robert Alton Harris in April of 1992. A total of five people were executed under his administration, the first two in the gas chamber, the latter three by lethal injection.

In Wilson's successful 1994 campaign for re-election against Kathleen Brown his two signature issues were his opposition to the billions spent by the State funding services for illegal immigrants and the race based quota components of Affirmative Action. Support for the overwhelmingly popular Prop 187 helped give him a landslide win.

Term limit laws passed by voters in Proposition 140, and championed by Wilson in 1990, prohibited the Governor from running for reelection to a third term, and Wilson left California with a sixteen billion dollar budget surplus.

Wilson also ran for President in the 1996 election, making major announcements on both the east and west coasts. He announced first at Battery Park in New York City with the Statue of Liberty as a backdrop, completed a cross-country tour, and made his west coast announcement at the Los Angeles Police Academy. The Wilson campaign had problems from the start, however. After deciding to run he almost immediately had throat surgery, which prevented him from speaking for months. His run lasted a month and a day, and left him with a million dollars in campaign debt.

Wilson is currently a distinguished visiting fellow at the Hoover Institution, a conservative think tank affiliated with Stanford University, the Ronald Reagan Presidential Foundation, the Richard M. Nixon Foundation, and the Donald Bren Foundation. He is also the founding director of the California Mentor Foundation and Chairman of the Board of Trustees of the National World War II Museum. Wilson sits on two prestigious Federal advisory committees, the President's Foreign Intelligence Advisory Board and the Defense Policy Board Advisory Committee, and currently works as a consultant at the Los Angeles office of Bingham McCutchen LLP, a large national law firm.

During and after Wilson's distinguished career he was awarded numerous awards and honors, including the Woodrow Wilson Institute Award for Distinguished Public Service, the Patriots Award by the Congressional Medal of Honor Society, an honorary degree from the San Diego State College of Professional Studies and Fine Arts, the Distinguished Alumnus Award from Boalt Hall, UC Berkeley, and the Bernard E. Witkin Amicus Curiae Award given by the Judicial Council of California.

In 2007 a statue of Pete Wilson joined Ernest Hahn and Alonzo Horton on what has been dubbed the "San Diego Walk of Fame." At the unveiling Wilson quipped, "View this statue, as I will, as a surrogate recipient of the tribute that's deserved by all of you who shared the dream, who made it come true and gave all the proud neighborhoods of San Diego the vibrant heart they needed."

PUBLIC SERVICE

JAMES BAKER
Secretary of State and Treasury

James Addison Baker served as the Chief of Staff in President Ronald Reagan's first administration and in the final year of the administration of President George H. W. Bush. He also served as Secretary of the Treasury in the second Reagan administration, and Secretary of State during the George H. W. Bush administration.

Baker was born in Houston, Texas on April 28, 1930 to James A. Baker, Jr. and Ethel Bonner (nee Means) Baker. His father was a partner the Houston law firm of Baker Botts. Baker attended the Hill School, a boarding school in Pottstown, Pennsylvania, and graduated from Princeton University in 1952. Afterwards he earned a J.D. from the University of Texas at Austin and began to practice law in Texas in 1957. From 1957 to 1969, and then from 1973 to 1975, he practiced law at the firm of Andrews & Kurth.

Baker served in the Marine Corps from 1952 to 1954, attained the rank of First Lieutenant, and later rose to Captain in the Marine Corps Reserve.

His first wife, the former Mary Stuart McHenry, was active in the Republican Party, working on the Congressional campaigns of George H. W. Bush. Originally Baker had been a Democrat, although he had been too busy trying to succeed in a competitive law firm to worry about politics and considered himself apolitical, but his wife's influence and enthusiasm led Baker to both politics as a career and to the

Republican Party. He was a regular tennis partner with Bush at the Houston Country Club in the late 1950s, and when Bush decided to run for the U.S. Senate in 1969 he supported Baker's decision to run for the Congressional seat he was vacating. Baker changed his mind about running however, when his wife was diagnosed with cancer. She died of breast cancer in February of 1970.

Bush then encouraged Baker to become active in politics to deal with the grief, something Bush had done when daughter Pauline Robinson (1949–1953) died of leukemia. Baker became chairman of Bush's Senate campaign in Harris County and although Bush lost to Lloyd Bentsen in the election, Baker continued in politics and became the Finance Chairman of the Republican Party in 1971. The following year he was selected as the Gulf Coast Regional Chairman for the Richard Nixon presidential campaign, but in 1973 and 1974 Baker returned to the full time practice of law at Andrews & Kurth.

Baker served as Undersecretary of Commerce under President Gerald Ford in 1975 and ran Ford's unsuccessful re-election campaign in 1976, and in 1978 ran unsuccessfully to become Attorney General of Texas, losing the election to future Governor Mark White. After serving as George H.W. Bush's campaign manager in the 1980 Republican primaries, Baker was named White House Chief of Staff by President Ronald Reagan in 1981 and served in that capacity until 1985.

Baker managed the President's 1984 re-election campaign in which Reagan won with a record total of 525 electoral votes (of 538 possible), and received 58.8% of the popular vote to Walter Mondale's 40.6%. In the new administration Baker "switched roles" with Secretary of the Treasury Donald Regan – a fellow Marine – who replaced Baker as

Chief of Staff. While serving as Treasury Secretary he then organized the Plaza Accord of September 1985 and the Baker Plan to target international debt. Baker also served on Reagan's National Security Council and remained Treasury Secretary through 1988, during which time he also served as campaign chairman for Bush's successful presidential bid.

President George H.W. Bush appointed Baker Secretary of State in 1989, and he served in that role through 1992. Then from 1992 to 1993 he served as Bush's White House Chief of Staff, the same position he had held from 1981 to 1985 during the first Reagan administration.

On January 9, 1991, during the Geneva Peace Conference with Iraqi Foreign Minister Tariq Aziz in Geneva, Secretary of State Baker declared, "If there is any use (of chemical or biological weapons), our objectives won't just be the liberation of Kuwait, but the elimination of the current Iraqi regime...." Baker later acknowledged the intent of this statement was to threaten a retaliatory nuclear strike on Iraq, and the Iraqis received his message when he helped to construct the thirty-four nation alliance which fought alongside the United States in the Gulf War.

James Baker was awarded the Presidential Medal of Freedom in 1991, and is the namesake of the James A. Baker III Institute for Public Policy at Rice University in Houston, Texas.

ROBERT BORK
Federal Judge and Supreme Court Nominee

Robert Heron Bork is an American legal scholar who has advocated the judicial philosophy of originalism. Bork formerly served as Solicitor General, acting Attorney General, and judge for the United States Court of Appeals for the District of Columbia Circuit. In 1987 he was nominated to the Supreme Court by President Ronald Reagan, but the Senate rejected his nomination. Bork had more success as an antitrust scholar, where his once-idiosyncratic view that antitrust law should focus on maximizing consumer welfare has come to dominate American legal thinking on the subject. He is currently a lawyer, law professor, and bestselling author.

Robert Bork was born on March 1, 1927 in Pittsburgh, Pennsylvania. His father was Harry Philip Bork, a steel company purchasing agent, and his mother Elisabeth Kunkle was a schoolteacher. Bork attended the Hotchkiss School in Lakeville, Connecticut and earned Bachelor's and law degrees from the University of Chicago, where he served on Law Review. At UC he was awarded a Phi Beta Kappa key with his law degree in 1953, and passed the bar in Illinois that same year. After a period of service in the Marine Corps, Bork began as a lawyer in private practice in 1954 and then was a professor at Yale Law School from 1962 to 1975 and 1977 to 1981. Among his students during this time were future President Bill Clinton, along with U.S. Senator and United States Secretary of State Hillary Rodham Clinton, Anita Hill, Robert Reich, Jerry Brown and John R. Bolton. Bork is best known for his theory that the only way to reconcile the role of the judiciary in American government against what he terms the "Madisonian" or "counter-majoritarian" dilemma of the judiciary - i.e. making law without popular approval - is for constitutional adjudication to be guided by the framers' original understanding of the Constitution. Reiterating that it is a court's task to adjudicate and not to "legislate from the bench," he has advocated that judges exercise restraint in deciding cases, emphasizing the role of the courts is to frame "neutral principles" (a term borrowed from Herbert Wechsler) and not simply ad hoc pronouncements or subjective value judgments. Bork once said, "The truth is the judge who looks outside the Constitution always looks inside himself and nowhere else."

Bork built on the influential critiques of the Warren Court authored by Alexander Bickel, who criticized the Supreme Court under Warren for shoddy and inconsistent reasoning, undue activism, and misuse of historical materials. Bork's critique was harder-edged than Bickel's however, and he has written, "We are increasingly governed not by law or elected representatives but by an unelected, unrepresentative, unaccountable committee of lawyers applying no will but their own." Bork's writings have influenced the opinions of conservative judges such as Supreme Court Associate Justice Antonin Scalia and former Chief Justice William Rehnquist, and have sparked a vigorous debate within legal academia about how the Constitution is to be interpreted.

At Yale Bork was best known for writing *The Antitrust Paradox*, a book in which he argued that consumers were often beneficiaries of corporate mergers, and that many then-current readings of the antitrust laws were economically irrational and hurt consumers. Bork's writings on antitrust law, along with those of Richard Posner and other law and economics thinkers, were heavily influential in causing a shift in the U.S. Supreme Court's approach to antitrust laws since the 1970s.

Bork served as Solicitor General in the U.S. Department of Justice from June 1973 to 1977, and in that capacity argued several high profile cases before the Supreme Court in the 1970s including 1974's *Milliken v. Bradley*, where Bork's brief in support of the State of Michigan was influential among the justices. Chief Justice Warren Burger called Bork the most effective counsel to appear before the Court during his tenure, and Bork hired many young attorneys as assistants who went on to have remarkable careers including Judges Danny Boggs and Frank H. Easterbrook - as well as Robert Reich, later to become President Bill Clinton's Secretary of Labor.

On October 20, 1973 Solicitor General Bork was instrumental in the "Saturday Night Massacre," President Richard Nixon's firing of Watergate Special Prosecutor Archibald Cox following Cox's request for tapes of Oval Office conversations. Nixon initially ordered Attorney General Elliot Richardson to fire Cox, but Richardson resigned rather than carry out the order. Deputy Attorney General William Ruckelshaus considered the order "fundamentally wrong" and also resigned, making Bork acting Attorney General. Although Bork believed Nixon's order to be valid and appropriate, he considered resigning to avoid being "perceived as a man who did the President's bidding to save my job." Richardson and Ruckelshaus told Bork he should not resign to avoid the damage a chain of resignations would do to the Department of Justice, and when Nixon reiterated his order Bork complied and fired Cox. He then remained acting Attorney General until the appointment of William B. Saxbe in mid-December.

Bork was later a circuit judge for the United States Court of Appeals for the District of Columbia Circuit between 1982 and 1988. He was nominated by President Reagan, confirmed by the Senate, and received his commission on February 9, 1982. One of his most significant opinions while on the D.C. Circuit was *Dronenburg v. Zech,* which was decided in 1984. This case involved James L. Dronenburg, a sailor who had been administratively discharged from the Navy for engaging in homosexual conduct. Dronenburg argued that his discharge violated his right to privacy, but this argument was rejected in an opinion by Bork in which he criticized cases in which the Supreme Court had enunciated a right to privacy.

Supreme Court Justice Lewis Powell was a moderate, and even before his expected retirement on June 27, 1987 Senate Democrats had asked liberal leaders to form "a solid phalanx" to oppose whomever President Ronald Reagan nominated to replace him - the assumption being it would tilt the court rightward. Democrats warned Reagan there would be a fight.

Reagan nominated Bork for the seat on July 1, 1987, and within forty-five minutes Ted Kennedy took to the Senate floor with a strong condemnation of Bork in a nationally televised speech in which he declared, "Robert Bork's America is a land in which women would be forced into back-alley abortions, blacks would sit at segregated lunch counters, rogue police could break down citizens' doors in midnight raids, schoolchildren could not be taught about evolution, writers and artists could be censored at the whim of the Government, and the doors of the Federal courts would be shut on the fingers of millions of citizens for whom the judiciary is - and is often the only - protector of the individual rights that are the heart of our democracy... President Reagan is still our President, but he should not be able to reach out from the muck of Irangate, reach into the muck of Watergate, and impose his reactionary vision of the Constitution on the Supreme Court and the next generation of Americans. No justice would be better than this injustice."

Bork complained, "There was not a line in that speech that was accurate," and in its obituary of Kennedy *The Economist* remarked that Bork was correct about the inaccuracy of Kennedy's speech, "but it worked." A brief was prepared for Joe Biden, head of the Senate Judiciary Committee, called the Biden Report. Bork contended in his best-selling book *The Tempting of America* that the report "so thoroughly misrepresented a plain record that it easily qualifies as world class in the category of scurrility."

TV ads narrated by Gregory Peck attacked Bork as an extremist, even as Kennedy's speech successfully fueled widespread public skepticism of Bork's nomination. The rapid response to Kennedy's "Robert Bork's America" speech stunned the Reagan White House, and the accusations went unanswered for two and a half months.

A hotly contested Senate debate over Bork's nomination ensued, partly fueled by the strong opposition of civil and women's rights groups concerned with what they claimed was Bork's desire to roll back civil rights decisions of the Warren and Burger courts. Bork is one of only three Supreme Court nominees to ever be opposed by the American Civil Liberties Union, along with William Rehnquist and Samuel Alito. He was also criticized for being an "advocate of disproportionate powers for the executive branch of Government, almost executive supremacy," which his role in the Saturday Night Massacre exemplified according to his critics.

During debate over his nomination Bork's video rental history was even leaked to the press. His video rental history was unremarkable, and included such harmless titles as *A Day at the Races, Ruthless People,* and *The Man Who Knew Too Much.* Writer Michael Dolan, who obtained a copy of the hand-written list of rentals, wrote waggishly about it for the *Washington City Paper.* Dolan justified accessing the list on the grounds that Bork himself had stated Americans only had such privacy rights as afforded them by direct legislation. The incident led to the enactment of the 1988 Video Privacy Protection Act.

To pro-life legal groups, Bork's originalist views and his belief that the Constitution does not contain a general "right to privacy" were viewed as a clear signal that, should he become a Justice on the Supreme Court, he would vote to reverse the Court's 1973 landmark decision *Roe v. Wade.* Accordingly, a large number of groups mobilized to press for Bork's rejection, and the resulting 1987 Senate confirmation hearings became an intensely partisan battle. Bork was faulted for his bluntness before the committee, including his criticism of the reasoning underlying *Roe v. Wade.* On October 23, 1987 the Senate rejected Bork's confirmation, with forty-two Senators voting in favor and fifty-eight voting against. Two Democratic Senators, David Boren and Ernest Hollings, voted in his favor, with six Republican Senators - including former Marines John Chafee and John Warner - voting nay. The vacant seat on the court to which Bork was nominated eventually went to Judge Anthony Kennedy and Bork, unhappy with his treatment in the nomination process, resigned his appellate-court judgeship in 1988.

According to columnist William Safire, the first published use of "bork" as a verb was possibly in *The Atlanta Journal-Constitution* on August 20, 1987. Safire defines "to bork" by reference "to the way Democrats savaged Ronald Reagan's nominee, Appeals Court judge Robert H. Bork, the year before." Perhaps the best known use of the verb to "bork" occurred in July of 1991 at a conference of the National Organization for Women in New York City when feminist Florynce Kennedy addressed the conference on the importance of defeating the nomination of Clarence Thomas to the Supreme Court. She said, "We're going to bork him. We're going to kill him politically... this little creep, where did he come from?" Thomas was subsequently confirmed after one of the most divisive confirmation fights in Supreme Court history.

In March of 2002 the *Oxford English Dictionary* added an entry for the verb Bork as U.S. political slang, with this definition: "To defame or vilify (a person) systematically, esp. in the mass media, usually with the aim of preventing his or her appointment to public office; to obstruct or thwart (a person) in this way." This usage appears unrelated to the computer slang term "borked" or "borken," which is a deliberate typo for "broken."

Following his resignation from the U.S. Court of Appeals for the D.C. Circuit Bork was for several years a senior fellow at the American Enterprise Institute for Public Policy Research, a conservative think tank. He has also consulted for Netscape in the Microsoft litigation, is currently a fellow at the Hudson Institute, and served as a visiting professor at the University of Richmond School of Law and as a professor at Ave Maria School of Law in Naples, Florida.

Bork has written several books, including the two best-sellers *The Tempting of America*, about his judicial philosophy and his nomination battle, and *Slouching Towards Gomorrah: Modern Liberalism and American Decline,* in which he argues that the rise of the New Left in the 1960s has undermined the moral standards necessary for civil society and spawned a generation of intellectuals who oppose Western civilization.

In 2003 he published *Coercing Virtue: The Worldwide Rule Of Judges,* an American Enterprise Institute book which includes his philosophical objections to the alleged phenomenon of incorporating international ethical and legal guidelines into the fabric of domestic law. In particular, he focuses on the problems he sees as inherent in the federal judiciary of three nations, Israel, Canada, and the United States, countries where he believes the courts have grossly exceeded their discretionary powers, and which have discarded precedent and common law and in their place substituted their own liberal judgment.

Bork also advocates a modification to the Constitution which would allow Congressional super-majorities to override Supreme Court decisions, similar to the Canadian notwithstanding clause. Although Bork has many moderate critics, some of his arguments have earned criticism from conservatives as well. While an opponent of gun control, Bork has denounced what he calls the "NRA view" of the Second Amendment, something he describes as the "belief that the Constitution guarantees a right to Teflon-coated bullets." Instead, he has argued that the Second Amendment merely guarantees a right to participate in a government militia.

A 2008 issue of the Harvard Journal of Law and Public Policy collected essays in tribute to Bork. Authors included Frank H. Easterbrook, George Priest, and Douglas Ginsburg.

WILLIAM A. EDDY
Minister to Saudi Arabia

William Alfred "Bill" Eddy (March 9, 1896 - May 3, 1962) was a U.S. Minister to Saudi Arabia, university professor, college president, intelligence officer and a Marine Corps officer who served in World War I and World War II.

Eddy was born in 1896 in the city of Sidon, which was at the time a part of Syria but is now in Lebanon. His parents, William King Eddy and Elizabeth Mills (nee Nelson) Eddy, were Presbyterian missionaries from the United States. Eddy grew up speaking English at home and in school, and Arabic on the streets with his friends. He stayed in the Middle East until high school, and then went to the College of Wooster for his college preparatory education. His overseas upbringing and firsthand knowledge of Arabic and the Arab culture would play a pivotal role in his life and in American–Saudi relations.

Following his graduation from Princeton University in 1917 and subsequent marriage to Mary Garvin, Eddy was accepted into the Marine Corps on June 6, 1917 as a "temporary Second Lieutenant" and was part of the first American Marine unit fighting in Europe in World War I, serving as an intelligence officer with the 6th Marine Regiment.

During the war he fought alongside other Marines during the German Offensive of 1918 and in the Battle of Belleau Wood against German Empire troops that same year. The battle is seen as an important success for allied forces against the Germans, and has since become part of Marine Corps folklore. Lieutenant Eddy was seriously wounded at Belleau Wood and was sent back to the U.S. For his actions as a Marine in World War I he received the Navy Cross, the Distinguished Service Cross, two Silver Stars, and two Purple Hearts.

After World War I Eddy taught at Peekskill Military Academy in New York, and in 1922 he received his Doctorate from Princeton University. He then embarked upon an academic career as a literary scholar and professor of English. In 1923 he was appointed the chair of the English Department at the American University in Cairo Egypt, but his wife and children found life in Egypt difficult and he later returned to the U.S. Then in 1928 he accepted a teaching position at Dartmouth College, and later served as president of both Hobart College and William Smith College.

With the threat of another World War looming, Eddy returned to the Marine Corps at the rank of Lieutenant Colonel and in 1941 became the Naval Attaché and Naval Attaché for Air in Cairo. He would work with both Naval Intelligence and the Office of Strategic Services (OSS) for the duration of the war, and was later to be a key figure in the formation of the CIA.

Early in the war Eddy suggested the United States try to become closer to Saudi Arabia because of its strategic importance and because of the country's relative independence and internal stability. In December of 1941 he was redeployed as Naval Attaché to Tangier, Morocco in order to try to help secure areas of North Africa under threat by the Germans, and was instrumental in obtaining intelligence there and in setting up an intelligence network which streamlined the process of conveying information from the field back to the U.S.

While in Tangier Eddy was also part of a group which helped organize subversive fighting elements in Spanish Morocco in case the Germans made it west. His intelligence work on the ground was a key to the success of allied operations, including the pivotal Patton-led Operation Torch which began in 1942.

In 1943 the Navy and the OSS agreed to cooperate in sending Eddy to Saudi Arabia as a U.S. State Department employee. His official title was "Special Assistant to the American Minister," and although he was resident at the American Legation in the city of Jeddah he was told to visit neighboring Gulf States as well in order to begin building the U.S.-Middle East relationships which were already beginning to emerge.

At the time President Franklin D. Roosevelt had already begun in earnest to begin a relationship with Saudi Arabia, and oil exploration and drilling was building up via the U.S. company CASOC, Aramco's predecessor. In 1944 Eddy met King Abdul-Aziz Al Saud (Ibn Saud) for the first time, and they would maintain a close relationship until the King's death in 1953. On September 23, 1944 Eddy became the "Envoy Extraordinary and Ministory Plenipotentiary," to Saudi Arabia, and remained in that post until May 28, 1946 - the second resident U.S. chief of mission to Saudi Arabia.

On February 14, 1945 King Abdul-Aziz had a historic meeting with President Roosevelt aboard the U.S. Naval ship *USS Quincy* on the Great Bitter Lake of the Suez Canal in Egypt, and Colonel Eddy was asked by the King to be translator for both himself and President Roosevelt during their conversation. It was the first time the King had left

Saudi Arabia, and much of the conversation was recorded by Eddy and used in a later work titled *FDR Meets Ibn Saud.*

On August 1, 1946 Eddy was appointed to the post of Special Assistant to the Secretary of State for Research and Intelligence, and was an instrumental figure in the passing of the National Security Act of 1947 which in essence allowed for the creation of the Central Intelligence Agency. Eddy and his family moved to Washington, D.C., where he worked on creating and developing the CIA as well as continuing the development of U.S.-Middle Eastern relations.

Throughout the 1950s Eddy worked as a consultant for the Arabian American Oil Company (Aramco) and was an instrumental figure in keeping Aramco–Saudi relations as peaceful as could be. Given King Abdul-Aziz's relationship with him and other Americans like Thomas Barger, the Saudis resisted completely nationalizing the company and instead brokered several key deals which would maintain American involvement while at the same time expanding the benefits brought by oil revenue to more Saudis.

William Eddy's final days were spent in Beirut, and he died on May 3, 1962 at the age of sixty-six in the hospital of the American University of Beirut which his father and family friends had helped to found. He was buried in the city of his birth, Sidon, in an Arab Christian graveyard, and his grave there is inscribed: "William Alfred Eddy. Colonel, U.S.M.C. Born Sidon, March 9, 1896. Died Beirut, May 3, 1962."

In 2008 the first biography of William Eddy, *Arabian Knight: Colonel Bill Eddy USMC and the Rise of American Power in the Middle East*, was published by Selwa Press. It was written by the Middle East specialist, author and *Washington Post* journalist Thomas Lippman.

SMITH HEMPSTONE
Ambassador to Kenya

Smith Hempstone Jr. (Feb 1, 1929 - Nov 19, 2006) was a journalist, author, the United States Ambassador to Kenya from 1989 to 1993, and a vocal proponent of democracy who fought for free elections there.

Hempstone was born in Washington, D.C., where his maternal grandfather and great-grandfather had been part-owners of *The Star.* He attended George Washington University, graduated from the University of the South, and won a Nieman fellowship to Harvard. He was also a U.S. Marine in the Korean War from 1950 through 1952, and left the Corps as a Captain.

After Korea Hempstone did radio rewrite for the *Associated Press* in Charlotte, North Carolina, was a reporter at the *Louisville Times* in Kentucky, worked as a rewrite man at *National Geographic* in Washington, D.C., and then as a reporter at the *Washington Star.* He was a fellow of the Institute of Current World Affairs in Africa, as well as a foreign correspondent for the *Chicago Daily News* in Africa and Latin America. He was also a foreign correspondent for the *Washington Star* in Latin America and Europe, and associate editor and editorial page director of the *Star.* He left the *Star* in 1975 after a disagreement with Joe L. Allbritton, its new owner, and began writing a syndicated twice-weekly column, *Our Times*, beginning in 1975.

While working in Africa as a correspondent for the *Chicago Daily News* Hempstone wrote several books, and also wrote a syndicated column carried by ninety newspapers. Then in 1982 he began working as editor of *The Washington Times*, and served as editor-in-chief from 1984 to 1985. While sitting on a Maine beach in 1987, the idea that he would be a fine Ambassador to Kenya popped into Hempstone's mind. He approached George H. W. Bush before his election the next year and said he would like the post, and President Bush remembered. He was appointed Ambassador to Kenya in 1989, at a time when the United States was beginning to push African countries toward democracy and human rights. Hempstone worked toward these goals by fighting for multi-party elections in Kenya in 1991, nine years after Kenyan president Daniel arap Moi had banned all parties except his own. The administration derided him, saying he failed to understand that a strong, unified government was necessary to keep Kenya's tribal groups from splitting the country. Hempstone responded by threatening to have aid cut unless democracy blossomed, all while castigating economic corruption, helping a leading dissident flee, and inviting government opponents to Embassy parties. "We don't just export Coca-Cola and blue jeans," he once said. "We export democracy." This caused the African press to describe his style as "bulldozer diplomacy." The official newspaper ran front-page cartoons depicting him as a fat pig, and one headline said, "Shut up, Mr. Ambassador."

Beyond carrying a gun in case of assassination attempts, Hempstone told *National Public Radio* that he had learned to take canapés from the back of the tray at diplomatic receptions on the theory that those would be less likely to be poisoned. The Kenyan government did in fact isolate him, and according to Hempstone in his book *Rogue Ambassador:*

217

An African Memoir, twice attempted to kill him. Multi-party elections were ultimately held in 1992, and were won by Moi with just thirty-six percent of the vote.

Bashfulness was never a problem for Hempstone. While on his honeymoon in Venice in 1954 he knocked on the door of a hotel suite occupied by Ernest Hemingway, who was only too pleased to converse with the stranger. "Been to Africa?" Hemingway asked. "You ought to go. Africa's man's country - fish, hunt, write - the best."

Many have suggested that Hempstone found a role model in Hemingway in more ways than following his advice about Africa. Beyond commonalities like girth, a white beard and a taste for tobacco and good liquor, Hempstone affected a Hemingwayesque style, reflected by the .38-caliber pistol he packed.

"He often acts more like a swashbuckling novelist than a diplomat," an article in the *Washington Post* said in 1993. In 1990 Hempstone had already responded to similar criticism in an article in the *New York Times.* "I'm not a diplomat," he said. The article noted that he chuckled.

Smith Hempstone died in 2006 at Suburban Hospital in Bethesda, Maryland from complications of diabetes at the age of seventy-seven.

JAMES L. JONES
National Security Advisor

James Logan Jones Jr. is the current United States National Security Advisor and a retired Marine Corps four-star General. During his military career he served as Commander, United States European Command (COMUSEUCOM) and Supreme Allied Commander Europe (SACEUR) from 2003 to 2006, and as the 32nd Commandant of the Marine Corps from July 1999 to January 2003. Jones retired from the Marine Corps on February 1, 2007 after forty years of service.

After retiring from the Marine Corps Jones remained involved in national security and foreign policy issues. In 2007 he served as the chairman of the Congressional Independent Commission on the Security Forces of Iraq, which investigated the capabilities of the Iraqi police and armed forces, and in November of 2007 he was appointed by the U.S. Secretary of State as special envoy for Middle East security. He also served as chairman of the Atlantic Council of the United States from June 2007 to January 2009, when he assumed the post of National Security Advisor.

Jones was born on December 19, 1943 in Kansas City, Missouri. He is the son of James L. Jones, Sr., a decorated Marine in World War II who was an officer in the Observer Group and the commanding officer of its successor, the Amphibious Reconnaissance Battalion. After spending his formative years in France, where he attended the American

School of Paris, Jones returned to the United States to attend the Georgetown University Edmund A. Walsh School of Foreign Service, from which he received a Bachelor of Science degree in 1966. While there Jones, who is six feet four inches tall, played forward on the Georgetown Hoyas men's basketball team.

In January of 1967 Jones was commissioned a Second Lieutenant in the U.S. Marine Corps, and upon completion of The Basic School at Marine Corps Base Quantico, Virginia in October of 1967 he was ordered to the Republic of Vietnam where he served as a platoon and company commander with Golf Company, 2nd Battalion, 3rd Marines. While overseas he was promoted to First Lieutenant in June of 1968.

Returning to the United States in December of 1968, Jones was assigned to Marine Corps Base Camp Pendleton, California where he served as a company commander until May of 1970. He then received orders to Marine Barracks, Washington, D.C. for duties as a company commander, and served in this assignment until July of 1973. While at this post he was promoted to Captain, and from July 1973 until June 1974 he was a student at the Amphibious Warfare School at the Marine Corps University in Quantico.

Following a tour in Okinawa, Japan Jones was assigned to Headquarters Marine Corps in Washington, D.C. During this assignment he was promoted to Major, and his next assignment was as the Marine Corps liaison officer to the United States Senate, where he served until July 1984. In this assignment, his first commander was John McCain, then a U.S. Navy Captain. He was promoted to Lieutenant Colonel in September of 1982.

After attending the National War College in Washington, D.C. Jones was assigned to command the 3rd Battalion, 9th

Marines, at Camp Pendleton, California from July 1985 to July 1987. In August of 1987 he returned to Headquarters Marine Corps, where he served as senior aide to the Commandant of the Marine Corps. He was promoted to Colonel in April of 1988, and became the military secretary to the Commandant in February of 1989. During August of 1990 he was assigned as the commanding officer of the 24th Marine Expeditionary Unit (24th MEU) at Camp Lejeune and participated in Operation Provide Comfort in Northern Iraq and Turkey. He was advanced to Brigadier General in 1992 and assigned duties as deputy director, J-3, U.S European Command, Stuttgart, Germany. During this tour of duty he was reassigned as chief of staff, Joint Task Force Provide Promise, for operations in Bosnia-Herzegovina and the Republic of Macedonia.

Returning to the United States, he was advanced to the rank of Major General in 1994 and assigned as commanding general of the 2nd Marine Division at Camp Lejeune. Jones next served as Director, Expeditionary Warfare Division, Office of the Chief of Naval Operations during 1996, then as the Deputy Chief of Staff for Plans, Policies, and Operations, Headquarters Marine Corps, Washington, D.C. He was advanced to Lieutenant General in July of 1996, and his next assignment was as the military assistant to the Secretary of Defense.

On April 21, 1999 Jones was nominated for appointment to the grade of General and assignment as the 32nd Commandant of the Marine Corps. He was promoted to General on June 30, 1999, assumed the post on July 1, and served as Commandant until January 2003 when he turned over the reins to General Michael Hagee.

Among other innovations during his career as Marine Corps Commandant, Jones oversaw the Corps' development of MARPAT camouflage uniforms and the adoption of the Marine Corps Martial Arts Program (MCMAP). These replaced M81 Woodland uniforms and the LINE combat system, respectively.

Jones next assumed duties as the Commander of U.S. European Command in January of 2003, and as Supreme Allied Commander Europe the following day. He was the first Marine Corps General to serve as SACEUR/EUCOM commander.

The Marine Corps had only recently begun to take on a larger share of high-level assignments in the Department of Defense. As of December 2006 Jones was one of five serving Marine Corps four-star General Officers who outranked the current Commandant of the Marine Corps (General James T. Conway) in terms of seniority and time in grade - the others being Chairman of the Joint Chiefs of Staff Peter Pace, former Commandant Michael Hagee, Commander of U.S. Strategic Command James E. Cartwright, and Assistant Commandant Robert Magnus.

Jones was reported to have declined an opportunity to succeed General John Abizaid as commander of U.S. Central Command. He stepped down as SACEUR on December 4, 2006 and retired from the Marine Corps in February of 2007.

Following his retirement from the military Jones became president of the Institute for 21st Century Energy, an affiliate of the U.S. Chamber of Commerce, and served as chair of the board of directors of the Atlantic Council of the United States from June 2007 until January 2009 when he assumed the post of National Security Advisor. He also served as a member of the guiding coalition for the Project on National Security Reform, as well as chairman of the Independent Commission on Security Forces of Iraq.

The Marines Have Landed

Secretary of State Condoleezza Rice twice asked Jones to be Deputy Secretary of State after Robert Zoellick resigned, but he declined. Rice appointed Jones as a special envoy for Middle East security on November 28, 2007, and he worked with Israelis and Palestinians to strengthen security for both sides.

On December 1, 2008 President-elect Obama announced Jones as his selection for National Security Advisor. The pick surprised many people because, as Michael Crowley reported, "The two men didn't meet until Obama's foreign policy aide, Mark Lippert, arranged a 2005 sit-down, and as of this October Jones had only spoken to Obama twice."

Crowley speculated that Jones' record suggested he was "someone who, unencumbered by strong ideological leanings, can evaluate ideas dispassionately whether they come from left or right," and, "this is probably why Obama picked him." Jones was also picked because he is well-respected and likely to possess the skills to navigate the other prestigious and powerful cabinet members. "He does not appear to be a natural antagonist of anyone else on the team. Though he doesn't know Gates especially well, both men share long experience in the national security establishment (Gates was in the Air Force, and previously headed the CIA).

Former U.S. Secretary of Defense William Cohen, who hired Jones as his military assistant, is quoted as saying Jones has a placid demeanor and a "methodical approach to problems - he's able to view issues at both the strategic and tactical level."

While he was Commandant of the Marine Corps General Jones often signed his emails as "Rifleman," since he had served as an infantry officer.

RAYMOND W. KELLY
Police Commissioner of the City of New York

Raymond Walter Kelly is the current Commissioner of the New York City Police Department and the first person to hold the post for two nonconsecutive tenures. Kelly has spent thirty-one years in the NYPD, serving in twenty-five different commands and as Police Commissioner from 1992-1994 and 2002 to the present. During his tenure Kelly has held most ranks, but was never a Three Star Bureau Chief or the Chief of Department. He also did not hold the position of Deputy Commissioner, being promoted directly from Two Star Chief to First Deputy Commissioner in 1990. After his handling of the World Trade Center bombing in 1993 he was mentioned as a possible candidate for FBI Director, and after Kelly turned down the position Louis Freeh was appointed.

Kelly was born on September 4, 1941 on the Upper West Side of Manhattan to James F. Kelly, a milkman, and Elizabeth Kelly, a dressing-room checker at Macy's. He is the father of Greg Kelly, co-host of the local Fox morning television show *Good Day New York.* Greg currently holds the rank of Lieutenant Colonel in the Marine Corps Reserve, and during his nine-year active duty military service was an A/V 8-B Harrier jump jet pilot assigned to Marine Attack Squadron 211, the "Wake Island Avengers." Kelly amassed over 150 aircraft carrier landings and flew over Iraq in

224

Operation Southern Watch, enforcing the United Nations imposed No-Fly Zone.

Ray Kelly graduated from academic and athletic powerhouse Archbishop Molloy High School in 1959, and earned a B.B.A. from Manhattan College in 1961. He also holds a J.D. from the St. John's University School of Law, an LL.M. from the New York University School of Law, and an M.P.A. from the John F. Kennedy School of Government at Harvard University. He obtained all of his academic degrees while working for the New York City Police Department.

Kelly received his Commission as a Second Lieutenant in the Marine Corps in 1963, and in 1965 went to the Republic of Vietnam with the 2nd Battalion, 1st Marines. As a First Lieutenant in Vietnam Kelly led troops in battle for most of his twelve months in country, including participation in Operation Harvest Moon. While there he was interviewed near Phu Bai by a young reporter named Dan Rather, which was a foreshadowing of Kelly's youngest son Greg, who would be assigned as an embedded journalist with the 3rd Infantry Division and become the first television journalist to televise pictures from Baghdad during the invasion of Iraq in 2003 - almost forty years later. Upon returning to the U.S. Kelly joined the Reserves and retired as a Colonel in the Marine Corps Reserve after thirty years of service.

Kelly joined the New York City Police Department as a trainee in 1960, graduated first in his class from the Police Academy, and passed the Sergeant's test before spending single a day on the beat. Upon graduation from the Academy Kelly won the "Bloomingdale Trophy" for the highest general average in shooting and in academic and physical prowess, and over the course of his career he received a total of fifteen citations for meritorious service.

He was appointed First Deputy Commissioner in February of 1990, and promoted from a two-star Assistant Chief to the First Deputy position over several three-star Bureau Chiefs and the four-star Chief of the Department, Robert J. Johnston Jr. At the time Johnston was so powerful that the traditional hierarchy was altered so that he would report directly to the Police Commissioner rather than the First Deputy as had been called for under the former departmental structure. This was to prevent Johnston from having to report to his former subordinate, Kelly.

On October 16, 1992 Raymond Kelly was appointed the 37th Police Commissioner of the City of New York. He took over a Police department that was 11.5% black in a city with over 25% black population, and at 9 AM on his first full day as Commissioner he was on the "black-owned" radio station WLIB for forty minutes talking to host Art Whaley, as well as callers, about minority recruitment.

As Commissioner, Kelly saw a reduction in crime almost from the outset, and it continued to go down as he aggressively pursued quality of life issues such as the "squeegee men" which had become a sign of decay in the city. Kelly also led the city through even what was at the time the worst terrorist attack on U.S. soil, the 1993 World Trade Center bombing, as then-Mayor David Dinkins was in Osaka, Japan at the time.

After Mayor Dinkins was defeated in his bid for re-election, new Mayor Giuliani replaced Kelly with Boston Police Chief Bill Bratton and an effort was made by the new administration to minimize the effects of the crime reducing strategies which had been put in place.

Kelly has occasionally been the target of criticism, such as for the department's handling of protests surrounding the 2004 Republican National Convention, but he has also received praise for revamping the department into a world class counter-terrorism operation. Prior to September 11th, 2001 there were fewer than two dozen officers working on terrorism full time, and today there are over one thousand.

One of Kelly's notable innovations was stationing New York detectives in cities throughout the world following terrorist attacks to determine if there were any connections to New York City. In the cases of both the March 11, 2004 Madrid bombing and the London bombings on July 7 and 21, 2005, NYPD detectives were on the scene within a day to relay pertinent information back to New York.

Kelly also served as Director of the International Police Monitors of the Multinational Force in Haiti from October 1994 through March 1995. This U.S.-led force was responsible for ending human rights abuses and establishing an interim police force there. For his service in Haiti, President Bill Clinton awarded him a commendation for "exceptionally meritorious service." Kelly was also awarded the Commander's Award for Public Service by then Chairman of the Joint Chiefs of Staff General Shalikashvili.

From 1996 to 1998 Kelly was Under Secretary for Enforcement at the United States Department of the Treasury. At that post he supervised the Department's enforcement bureaus, including the Customs Service, the Secret Service and the Bureau of Alcohol, Tobacco and Firearms. He was also responsible for the Federal Law Enforcement Training Center, the Financial Crimes Enforcement Network, and the Office of Foreign Assets Control. Then from 1998 to 2001 Kelly served as the Commissioner of the U.S. Customs Service, where he managed the agency's twenty thousand employees and twenty billion annual dollar budget.

In 2003 the National Father's Day Committee named Ray Kelly "Father of the Year," and in March of 2006 he was named Irish American of the Year by *Irish American Magazine*. Then in June Kelly received the Hundred Year Association of New York's Gold Medal "in recognition of outstanding contributions to the City of New York," and in September of that same year he was awarded the Légion d'honneur during a ceremony at the French consulate in Manhattan presided over by President of France Nicolas Sarkozy.

ROBERT S. MUELLER III
Director of the FBI

Robert Swan Mueller III is the current Director of the United States Federal Bureau of Investigation.

Mueller was born on August 7, 1944 in New York City, the son of Robert Swan and Alice C. (née Truesdale) Mueller, and grew up outside of Philadelphia, Pennsylvania. A 1962 graduate of St. Paul's School, he went on to receive a B.A. from Princeton University in 1966, an M.A. in international relations from New York University in 1967, and a Juris Doctor from the University of Virginia School of Law.

Mueller then joined the Marine Corps, where he served as an infantry officer for three years and led a rifle platoon of the 3rd Marine Division during the Vietnam War. He is a recipient of the Bronze Star, two Navy Commendation Medals, the Purple Heart, and the Vietnamese Cross of Gallantry.

After his military service Mueller continued his studies at the University of Virginia Law School and eventually served on the Law Review. After receiving his law degree, he worked as a litigator in San Francisco until 1976. He then served for twelve years in United States Attorney offices, first working in the office of the U.S. Attorney for the Northern District of California in San Francisco, where he rose to be chief of the criminal division, and then in 1982 he moved to Boston to work in the U.S. Attorney's Office for the District of Massachusetts as Assistant United States Attorney, where he investigated and prosecuted major financial fraud, terrorism and public corruption cases, as well as narcotics conspiracies and international money launderers.

After serving as a partner at the Boston law firm of Hill and Barlow, Mueller was again called to public service when in 1989 he served in the United States Department of Justice as an assistant to Attorney General Dick Thornburgh. The following year he took charge of its criminal division, and during his tenure oversaw prosecutions which included Panamanian leader Manuel Noriega, the Pan Am Flight 103 (Lockerbie bombing) case, and Gambino crime family boss John Gotti. In 1991, he was elected a fellow of the American College of Trial Lawyers.

In 1993 Mueller became a partner at Boston's Hale and Dorr, and specialized in white-collar crime litigation. He returned to public service in 1995 as senior litigator in the homicide section of the District of Columbia United States Attorney's Office, and in 1998 was named U.S. Attorney for the Northern District of California, a position he held until 2001.

Mueller was nominated for the position of FBI Director on July 5, 2001. Confirmation hearings in front of the Senate Judiciary Committee were quickly set for July 30th, only three days before Mueller was to undergo prostate cancer surgery. The vote on the Senate floor was unanimous, with his nomination being approved 98-0. He then served as Acting Deputy Attorney General of the United States Department of Justice for several months before officially becoming the FBI Director on September 4, 2001 - just one week before the September 11 attacks.

Director Mueller, along with Acting Attorney General James B. Comey, offered to resign from office in March of

2004 if the White House overruled a Department of Justice finding that domestic wiretapping without a court warrant was unconstitutional. Attorney General John Ashcroft denied his consent to attempts by White House Chief of Staff Andrew Card and Counsel Alberto Gonzales to waive the Justice Department ruling and permit the domestic warrantless eavesdropping program to proceed, and on March 12, 2004 President George W. Bush gave his support to changes in the program sufficient to satisfy the concerns of Mueller, Ashcroft and Comey. The extent of the National Security Agency's domestic warrantless eavesdropping under the President's Surveillance Program is still largely unknown.

DONALD REGAN
Secretary of the Treasury

Donald Thomas Regan (Dec 21, 1918 - June 10, 2003) was the Secretary of the Treasury from 1981 to 1985, and Chief of Staff from 1985 to 1987 in the Ronald Reagan Administration where he advocated "Reaganomics" and tax cuts to create jobs and stimulate production. Although he was very effective, Regan was sometimes criticized for his Prime Ministerial style, his involvement in the Iran-Contra Affair, and his frequent disagreements with First Lady Nancy Reagan.

Born in Cambridge, Massachusetts and of Irish Catholic origins, Don Regan earned his Bachelor's degree in English from Harvard University in 1940 and attended Harvard Law School before dropping out to join the Marine Corps at the onset of World War II. He reached the rank of Lieutenant Colonel while serving in the Pacific theater, and was involved in five major campaigns including Guadalcanal and Okinawa.

After the war he joined Merrill Lynch & Co. in 1946 as an account executive trainee, worked his way up through the ranks, and eventually took over as Merrill Lynch's chairman and CEO in 1971 - the year the company went public. He held those positions until 1980.

Regan was one of the original directors of the Securities Investor Protection Corporation and was vice chairman of the New York Stock Exchange from 1973 to 1975. He was a major proponent of brokerage firms going public, which he viewed as an important step in the modernization of Wall Street, and under his supervision Merrill Lynch had its IPO on June 23, 1971 and became only the second Wall Street firm to go public, after Donaldson, Lufkin & Jenrette.

During his tenure Regan also pushed hard for an end to minimum fixed commissions for brokers, which were fees brokerage companies had to charge clients for every transaction they made on the clients' behalf. Regan saw them as a cartel-like restriction, and in large part thanks to his lobbying fixed commissions were abolished in 1975.

In 1981 President Ronald Reagan selected Regan to serve as Treasury Secretary and made him a spokesman for his economic policies, dubbed "Reaganomics." He helped engineer tax reform, reduce income tax rates and ease tax burdens on corporations.

Regan unexpectedly swapped jobs with White House Chief of Staff (and fellow Marine) James Baker in 1985 and became closely involved in the day to day management of White House policy. That led Howard Baker, his successor as Chief of Staff, to say Regan was becoming a "Prime Minister" inside an increasingly complex Presidency. Regan resigned from his post in 1987 due to his involvement with the Iran-Contra Affair and frequent clashes with the President's wife, First Lady Nancy Reagan. He was seen as the fall guy for the affair, and the tongue-in-cheek saying that "Reagan had Regan" echoed throughout Washington.

Don Regan retired quietly to Virginia with Ann, his wife of over sixty years, and late in life spent nearly ten hours a day in his art studio painting landscapes, some of which sold for thousands of dollars and hang in museums. He died in 2003 of heart failure in Williamsburg, Virginia at the age of eighty-four.

GEORGE SCHULTZ
Secretary of State, Labor, and the Treasury

George Pratt Shultz is an economist, statesman, and businessman who served as Secretary of Labor from 1969 to 1970, Secretary of the Treasury from 1972 to 1974, and Secretary of State from 1982 to 1989. Before entering politics, he was professor of economics at MIT and the University of Chicago, and served as Dean of the University of Chicago Graduate School of Business from 1962 to 1969. Between 1974 and 1982 Shultz was an executive at Bechtel, eventually becoming the firm's president, and he is currently a distinguished fellow at Stanford University's Hoover Institution.

Shultz was born on December 13, 1920 in New York City, the son of Birl Earl Shultz and Margaret Lennox Pratt. In 1938 he graduated from the Loomis Chaffee School in Windsor, Connecticut, and he attended college at Princeton University, majoring in economics with a minor in public and international affairs. His senior thesis was an examination of the Tennessee Valley Authority's effect on local agriculture, for which he conducted on-site research, and he graduated with honors in 1942.

Following his college graduation Shultz joined the Marine Corps and served until 1945, attaining the rank of Captain. Then in 1949 he earned a Ph.D. in industrial economics from the Massachusetts Institute of Technology.

Schultz taught in both the MIT Department of Economics and the MIT Sloan School of Management from 1948 to 1957, with a leave of absence in 1955 to serve on President Dwight Eisenhower's Council of Economic Advisers as a senior staff economist. Then in 1957 he joined the University of Chicago Graduate School of Business as professor of industrial relations, and was later named dean in 1962.

Shultz served as President Richard Nixon's Secretary of Labor from 1969 to 1970, during which time he forced Pennsylvania construction unions which refused to accept black members to admit a certain number by an enforced deadline in what was essentially the first use of racial quotas by the federal government. He then became the first Director of the Office of Management and Budget.

Schultz was United States Secretary of the Treasury from May 1972 to May 1974, and during his tenure he was concerned with two major issues - the continuing domestic administration of Nixon's "New Economic Policy," which had begun under Secretary John B. Connally, and a renewed dollar crisis that broke out in February of 1973.

Shultz's attention was increasingly diverted from the domestic economy to the international arena however, and he participated in an international monetary conference in Paris in 1973 which grew out of the 1971 decision to abolish the gold standard - a decision both Shultz and Paul Volcker had supported. The conference formally abolished the Bretton Woods system, thereby causing all currencies to float.

Shultz resigned shortly before Nixon to return to private life, and in 1974 he left government service to become president and director of Bechtel Group, a large engineering and services company, where he remained until July of 1982 when he was appointed by President Ronald Reagan to serve as the sixtieth Secretary of State, replacing Alexander Haig, who had resigned. Shultz would serve for six and a half years - the longest tenure since Dean Rusk.

Shultz relied primarily on the Foreign Service to formulate and implement Reagan's foreign policy. By the summer of 1985 he had personally selected most of the senior officials in the Department, emphasizing professional over political credentials. The Foreign Service responded in kind by giving Shultz its "complete support," and making him one of the most popular Secretaries since Dean Acheson.

When Mikhail Gorbachev came to power in 1985 Shultz advised President Reagan to pursue a personal dialogue with him. This relationship produced its most practical result in December of 1987, when the two leaders signed the Intermediate Range Nuclear Forces Treaty. The treaty, which eliminated an entire class of missiles in Europe, was a milestone in the history of the Cold War. Although Gorbachev took the initiative, Reagan was well prepared by the State Department to adopt a policy of negotiations.

In response to the escalating violence of the Lebanese civil war, Reagan sent a Marine contingent to protect the Palestinian refugee camps and support the Lebanese Government. The October 1983 bombing of the Marine barracks in Beirut killed 241 U.S. servicemen, after which the deployment came to an ignominious end. Shultz subsequently negotiated an agreement between Israel and Lebanon, and convinced Israel to begin a partial withdrawal of its troops in January 1985 despite Lebanon's contravention of the settlement.

During the First Intifada (Arab-Israeli conflict), Shultz "proposed... an international convention in April 1988... on an interim autonomy agreement for the West Bank and Gaza Strip, to be implemented as of October for a three-year period." By December of 1988, following six months of shuttle diplomacy, Shultz had established a diplomatic dialogue with the Palestine Liberation Organization, which was picked up by the next Administration.

Shultz was well known for outspoken opposition to the "arms for hostages" scandal that would eventually become the Iran Contra situation. In his 1983 testimony before Congress he said the Sandinista government in Nicaragua was "a cancer in our own land mass" that must be "cut out." He was also opposed to any negotiation with the government of Daniel Ortega, saying, "Negotiations are a euphemism for capitulation if the shadow of power is not cast across the bargaining table."

George Shultz left office on January 20, 1989 but continues to be a strategist for the Republican Party. He was an advisor for George W. Bush's presidential campaign during the 2000 election, and the senior member of the so-called "Vulcans," a group of policy mentors for Bush which also included among its members Dick Cheney, Paul Wolfowitz and Condoleezza Rice. One of his most senior advisors and confidants is former Ambassador Charles Hill, who holds dual positions at the Hoover Institution and Yale University. Shultz has been called the father of the "Bush Doctrine," because of his advocacy of preventive war. He generally defended the Bush administration's foreign policy.

After leaving public office in 1989 Shultz became the first prominent Republican to call for the legalization of recreational drugs. He went on to add his signature to an advertisement, published in *The New York Times* on June 8, 1998 entitled, "We believe the global war on drugs is now causing more harm than drug abuse itself."

He also has spoken out against the Cuban embargo, going as far as calling the U.S. policy towards Cuba "insane." He has argued that free trade would help bring down Fidel

Castro's regime, and that the embargo only helps justify the continued repression in the island.

Shultz is the chairman of the JP Morgan Chase Bank's International Advisory Council and an honorary director of the Institute for International Economics. He is a member of the Hoover Institution at Stanford University, the Washington Institute for Near East Policy (WINEP) Board of Advisors, the New Atlantic Initiative, the prestigious Mandalay Camp at the Bohemian Grove, the Committee for the Liberation of Iraq, and the Committee on the Present Danger. He is also honorary chairman of the Israel Democracy Institute. Shultz formerly served on the board of directors for the Bechtel Corporation, Charles Schwab Corporation and Gilead Sciences, and is currently a co-chairman of the North American Forum.

George Schultz met his future wife, Nurse Lieutenant Helena Maria "Obie" O'Brien (1915-1995), while serving with the Marines in Hawaii, and together they had five children. Then in 1997, after the death of Helena, he married Charlotte Mailliard Swig, a prominent San Francisco socialite. Their marriage was called the "Bay Area Wedding of the Year," and they remain a power couple in San Francisco.

BING WEST
Assistant Secretary of Defense

Francis J. "Bing" West is an author of military books who served as Assistant Secretary of Defense for International Security Affairs during the Reagan Administration. His 2004 book *The March Up: Taking Bagdhad with the First Marine Division*, written with Marine Corps General Ray L. Smith, was awarded the Marine Corps Heritage Prize for Nonfiction, as well as the Colby Award. He is the father of the model Kaki West and former Marine Owen West.

West was born in the Dorchester section of Boston on May 3, 1941. He is a graduate of Georgetown University and Princeton University, where he was a Woodrow Wilson Fellow. He served as an infantry officer in the Marine Corps during the Vietnam War, and commanded a Combined Action Platoon which fought for 485 days in a remote village. A Captain at the time, he survived an attack that killed nine of the fifteen Marines. He was also a member of the Force Reconnaissance team that initiated "Operation Stingray" - small unit attacks behind enemy lines.

West authored a study while a Visiting Research Associate at the RAND Corporation in the mid-1960s entitled *The Strike Teams: Tactical Performance and Strategic Potential*. This paper was presented at the Annual DoD Counter-insurgency Research and Development

Symposium in October of 1968, and the RAND Military Systems Simulations Group implemented a classified model of this concept. This doctrinal innovation was directly opposed by MACV (Military Assistance Command Vietnam) in favor of the Army's concept of Air-Mobility "Fire and Thunder Operations," and by way of rebuttal West wrote *The Village*, in which he chronicled the daily lives of fifteen Marines who protected the people by living amongst them in their hamlets. The book became a classic of practical counterinsurgency, and has been on the Commandant's Required Reading List for thirty-six years.

West served as Assistant Secretary of Defense for International Security Affairs in the Reagan administration, and is currently president of the GAMA Corporation, which designs wargames and combat decision-making simulations. He is also the author of several books relating to the United States Military, and his collaboration with retired Marine Major General Ray "E-Tool" Smith, *The March Up*, was awarded the Marine Corps Heritage prize for nonfiction, as well as the Colby award for military nonfiction. West also authored a foreword for *Boredom by Day, Death by Night: An Iraq War Journal* by Marine Sergeant Seth Conner.

Grunts have been close to West's heart since he was a kid of four or five in Dorchester, when a bunch of Marines home on leave from the battlegrounds of World War II would hang out in a large attic which West's father, a doctor, let them use as a kind of clubhouse. "My mother thought, 'Well, these are perfect babysitters,' so from the age of two on my babysitters were Marines just back from the war. That obviously had a terrific effect. I remember walking back and forth, marching, and having a great time. You learn some good values in the Marines," he joked, "to be kind, understanding, docile..."

West's great-uncle and two uncles were Marines, as was his son Owen, who trades natural gas on Wall Street for Goldman Sachs and is an author and adventure racer who has competed in three Eco-Challenges and was the lone male on Team Playboy Extreme. Owen first served in the Corps from 1991 to 1996 and has since been to Iraq three times as part of Operation Iraqi Freedom as a Marine infantry officer, once in 2003, and twice in 2007.

The nickname "Bing" was bestowed on young Francis by an aunt who was taken with the fact he was born on Bing Crosby's birthday, May 3rd, and over the years the name lost its quotation marks and hardened into the real thing. West is currently a correspondent for *The Atlantic Monthly*, appears on *The News Hour* on PBS, and is a member of the Council on Foreign Relations He is currently writing the screenplay for *No True Glory* for Universal Studios with his son Owen.

The Marines Have Landed

NASA

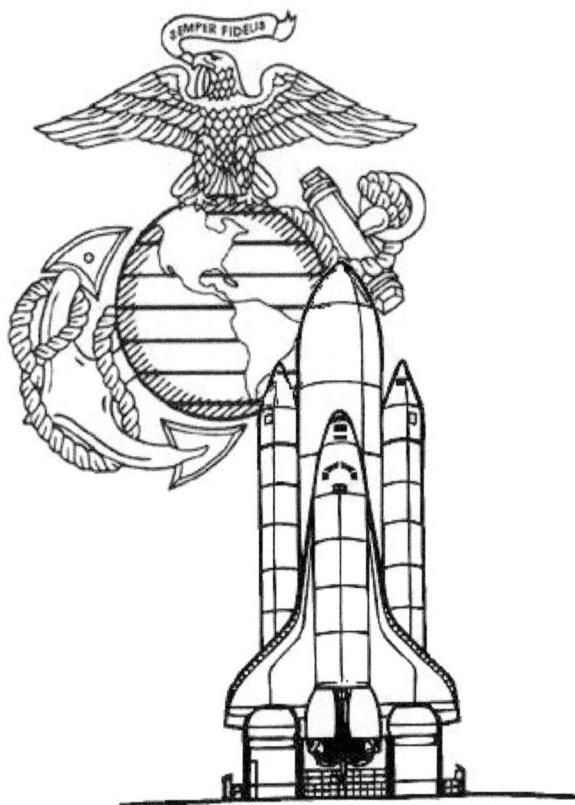

JOSEPH M. ACABA
Discovery

Joseph Michael "Joe" Acaba is an educator, hydrogeologist, and NASA astronaut. In May of 2004 he became the first person of Puerto Rican heritage to be named as a NASA astronaut candidate when he was selected as a member of NASA Astronaut Training Group 19. He completed his training on February 10, 2006 and was assigned to STS-119, which flew from March 15 to March 28, 2009 to deliver the final set of solar arrays to the International Space Station.

Acaba's parents, Ralph and Elsie Acabá, are from Hatillo, Puerto Rico. They moved in the mid-1960s to Inglewood, California where Joseph was born on May 17, 1967, and later relocated to Anaheim. Since his childhood Acaba has enjoyed reading, especially science fiction, and in school he excelled in both science and math. As a child his parents constantly exposed him to educational films, but it was the 8-mm film showing astronaut Neil Armstrong's Moon landing which really intrigued him about outer space. During his senior year in high school Acaba became interested in scuba diving and became a certified diver through a job training program at his school, and this experience inspired him to further his academic education in the field of geology after he graduated with honors from Esperanza High School in Anaheim in 1985. He received his Bachelor's degree in Geology from the University of California-Santa Barbara in

1990, and in 1992 earned his Master's degree in Geology from the University of Arizona.

Acaba was a Sergeant in the Marine Corps Reserve where he served for six years, and also spent two years in the Peace Corps where he taught modern teaching methodologies to three hundred teachers in the Dominican Republic. He then served as Island Manager of the Caribbean Marine Research at Lee Stocking Island in the Exumas, Bahamas, and upon his return to the U.S. he moved to Florida and became Shoreline Revegetation Coordinator in Vero Beach.

On May 6, 2004 Acaba and ten other people were selected by NASA as astronaut candidates from a pool of ninety-nine applicants. NASA administrator Sean O'Keefe, in the presence of Marine astronaut John Glenn, announced the members of the 19th group of Astronaut Candidates, an event which hadn't been repeated since 1958 when the original group of astronauts was presented to the world. Acaba, who was selected as an Educator Mission Specialist, completed his astronaut training in February of 2006 along with the other ten astronaut candidates and was assigned to the Hardware Integration Team in the International Space Station branch, where he worked on technical issues with European Space Agency (ESA) hardware.

Acaba was assigned to the crew of STS-119 as Mission Specialist Educator, which was launched on March 15, 2009 to deliver the final set of solar arrays to the International Space Station. Acaba, who carried on his person a Puerto Rican flag, requested that the crew be awakened on March 19 with the Puerto Rico folklore song "Qué Bonita Bandera" (What a Beautiful Flag), which refers to the Puerto Rican flag and was written in 1971 by Florencio Morales Ramos (Ramito) and sung by Jose Gonzalez and Banda Criolla.

On March 20 he provided support to the first mission spacewalk, and on March 21 he performed a spacewalk in which he helped to successfully unfurl the final "wings" of the solar array which will augment power to the ISS as it prepares to double its capacity to house six astronauts in the future. On March 28 the Space Shuttle *Discovery* and its crew of seven safely touched down at Kennedy Space Center in Florida, and afterward Acaba was amazed at the views from the space station, saying, "It was kind of surreal to look out the window and see your two buddies that you've been training with for a long time and see them out there... it was a special moment."

On March 18, 2008 Joseph Acaba was honored by the Senate of Puerto Rico, which sponsored his first trip to the Commonwealth of Puerto Rico since being selected for space flight. During his visit, which was announced by the President of the Puerto Rican Senate, the Honorable Kenneth McClintock, he met with schoolchildren at the Capitol as well as at the Bayamón, Puerto Rico Science Park, which includes a planetarium and several surplus NASA rockets among its exhibits. Acaba, returned to Puerto Rico once again on June 1, 2009, and during this visit was presented with a proclamation by Governor Luis Fortuño. He also received the Ana G. Mendez University System Presidential Medal and a Doctorate Honoris Causa from the Polytechnic University of Puerto Rico.

ANDREW M. ALLEN
Atlantis and Columbia

Andrew Michael "Andy" Allen is a retired Astronaut. A former United States Marine Corps aviator and Lieutenant Colonel, he worked as a test pilot before joining NASA in 1987, and flew three shuttle missions before retiring in 1997.

Allen was born on August 4th, 1955 in Philadelphia, Pennsylvania and graduated from Archbishop Wood Catholic High School in 1973.

He received his commission in the Marine Corps at Villanova University in 1977, and following graduation from flight school flew F-4 Phantoms from 1980 to 1983 with VMFA-312 at Marine Corps Air Station Beaufort, South Carolina, and was additionally assigned as the Aircraft Maintenance Officer. He was selected by Headquarters Marine Corps for fleet introduction of the F/A-18 Hornet, and was assigned to VMFA-531 at Marine Corps Air Station El Toro, California from 1983 to 1986. During his tour with VMFA-531 he was assigned as the squadron Operations Officer, and also attended and graduated from the Marine Weapons & Tactics Instructor Course, and the United States Navy Fighter Weapons School (Top Gun). A 1987 graduate of the United States Naval Test Pilot School at Naval Air Station Patuxent River, Maryland, he was a test pilot under instruction when advised of his selection to the astronaut program.

Over the course of his career Allen has logged over six thousand flight hours in more than thirty different aircraft. He has been awarded the Defense Superior Service Medal, Legion of Merit, Distinguished Flying Cross, Defense Meritorious Service Medal, Single Mission Air Medal, NASA Outstanding Leadership Medal, and NASA Exceptional Service Medal.

Allen retired from the Marine Corps and left NASA in October of 1997. He has since served in various industry leadership positions including President of the FIRST (For Inspiration and Recognition of Science and Technology) Foundation, Associate Program Manager for Ground Operations with United Space Alliance, and is currently a Vice President and Program Manager with Honeywell.

CHARLES F. BOLDEN, JR.
Columbia, Discovery, and Atlantis

Charles Frank "Charlie" Bolden, Jr. is the current Administrator of NASA, a retired United States Marine Corps Major General, and a former NASA astronaut. A 1968 graduate of the United States Naval Academy, he became a Marine Aviator and test pilot, and after his service as an astronaut became Deputy Commandant of Midshipmen at the Naval Academy. Bolden is the virtual host of the Shuttle Launch Experience attraction at Kennedy Space Center, and serves on the board of directors for the Military Child Education Coalition.

Bolden was born on August 19, 1946 in Columbia, South Carolina and graduated from C. A. Johnson High School in Columbia in 1964. He earned a Bachelor of Science degree in electrical science from the United States Naval Academy in 1968, and a Master of Science in systems management from the University of Southern California in 1977.

Bolden accepted a commission as a Second Lieutenant in the Marine Corps following graduation from the United States Naval Academy in 1968. He underwent flight training at Pensacola, Florida, Meridian, Mississippi and Kingsville, Texas before being designated a naval aviator in May of 1970. He flew more than one hundred sorties into North and South Vietnam, Laos, and Cambodia in the A-6A Intruder while assigned to VMA(AW)-533 at Royal Thai Air Base Nam Phong, Thailand from June 1972 to June 1973.

Upon his return to the United States Bolden began a two-year tour as a Marine Corps selection and recruiting officer in Los Angeles, California, followed by three years in various assignments at Marine Corps Air Station El Toro. In June of 1979 he graduated from the United States Naval Test Pilot School at Naval Air Station Patuxent River, Maryland and was assigned to the Naval Air Test Center's Systems Engineering and Strike Aircraft Test Directorates. While there he served as an ordnance test pilot and flew numerous test projects in the A-6E, EA-6B, and A-7C/E airplanes, logging more than six thousand hours flying time.

Bolden was selected as an astronaut candidate by NASA in 1980, and was a member of the Astronaut Corps until 1994 when he returned to active duty in the Marine Corps as the Deputy Commandant of Midshipmen at the Naval Academy. In July of 1997 he was assigned as the Deputy Commanding General of I Marine Expeditionary Force, and from February to June 1998 served as Commanding General, I MEF (FWD) in support of Operation Desert Thunder in Kuwait. In July of 1998 he was promoted to his final rank of Major General and assumed his duties as the Deputy Commander, United States Forces Japan. He then served as the Commanding General of the 3rd Marine Aircraft Wing from August 2000 until August 2002, and retired from the Corps in August of 2004.

First selected by NASA in May 1980, Bolden became an astronaut in August 1981. His many technical assignments included Astronaut Office Safety Officer, Technical Assistant to the Director of Flight Crew Operations, Special Assistant to the Director of the Johnson Space Center, Astronaut Office Liaison to the Safety, Reliability and Quality Assurance Directorates of the Marshall Space Flight Center and the Kennedy Space Center, Chief of the Safety

Division at JSC, Lead Astronaut for Vehicle Test and Checkout at the Kennedy Space Center, and Assistant Deputy Administrator, NASA Headquarters. A veteran of four space flights, he has logged over 680 hours in space. Bolden served as pilot on STS-61C in 1986 and STS-31 in 1990, and was the mission commander for STS-45 in 1992 and STS-60 in 1994.

Bolden was the first person to ride the Launch Complex 39 slidewire baskets which enable rapid escape from a space shuttle on the launch pad. The need for a human test was determined following a launch abort on STS-41-D, when controllers were afraid to order the crew to use the untested escape system.

On STS-61-C Bolden piloted Space Shuttle *Columbia.* During the six-day flight, crew members deployed the SATCOM KU satellite and conducted experiments in astrophysics and materials processing. STS-61-C launched from Kennedy Space Center in January of 1986 and the mission was accomplished in ninety-six orbits of the Earth, ending with a successful night landing at Edwards Air Force Base.

Bolden piloted Space Shuttle *Discovery* during STS-31. Launched in April of 1990 from Kennedy Space Center, the crew spent the five-day mission deploying the Hubble Space Telescope and conducting a variety of middeck experiments. They also used a variety of cameras, including both the IMAX in-cabin and cargo-bay cameras, for Earth observations from their record-setting altitude of over 400 miles. Following seventy-five orbits of Earth in 121 hours, *Discovery* landed at Edwards Air Force Base.

On STS-45 Bolden commanded a crew of seven aboard Space Shuttle *Atlantis,* launched in March of 1992. STS-45 was the first Spacelab mission dedicated to NASA's Mission to Planet Earth. During the nine-day mission the crew operated the twelve experiments which constituted the ATLAS-1 (Atmospheric Laboratory for Applications and Science) cargo. ATLAS-1 obtained a vast array of detailed measurements of atmospheric chemical and physical properties which contribute significantly to improving our understanding of our climate and atmosphere. In addition, this was the first time an artificial beam of electrons was used to stimulate a man-made auroral discharge. Following 143 orbits of Earth, Atlantis landed at Kennedy Space Center.

Bolden commanded STS-60's crew of six aboard *Discovery*. This was the historic first joint-American/Russian Space Shuttle mission involving the participation of Russian cosmonaut Sergei Krikalyov as a mission specialist crew member. The flight launched in February of 1994 from Kennedy Space Center and carried the Space Habitation Module-2 (SPACEHAB) and the Wake Shield Facility. Additionally, the crew conducted a series of joint U.S./Russian science activities. The mission achieved 130 orbits of the Earth, ending with a landing on at the Kennedy Space Center.

On May 23, 2009 President Barack Obama announced the nomination of Bolden as NASA Administrator. He was confirmed by the Senate on July 15, 2009, and is the first African American to head the agency on a permanent basis.

VANCE D. BRAND
Apollo, Columbia and Challenger

Vance DeVoe Brand is an engineer, former test pilot and NASA astronaut. He served as command module pilot during the first U.S.-Soviet joint space flight in 1975, and as commander of three space shuttle missions.

His flight experience includes 9,669 flying hours, with 8,089 hours in jets, 391 hours in helicopters, 746 hours in spacecraft, and checkout in more than thirty types of military aircraft.

Brand was born in Longmont, Colorado on May 9, 1931 and is the son of Rudolph William and Donna Mae Brand. He was active in Troop 64 of the Boy Scouts of America in Longmont, where he achieved its second highest rank, Life Scout. Brand graduated from Longmont High School and received a Bachelor of Science degree in Business from the University of Colorado at Boulder in 1953, a Bachelor of Science degree in Aeronautical Engineering from the same institution in 1960, and a Master's degree in Business Administration from UCLA in 1964.

Brand was commissioned as an officer and served as a naval aviator in the Marine Corps from 1953 to 1957. His military assignments included a fifteen-month tour in Japan as a jet fighter pilot, and following his release from active duty he continued service in Marine Forces Reserve and Air National Guard jet fighter squadrons until 1964.

Employed as a civilian by the Lockheed Corporation from 1960 to 1966, he worked initially as a flight test engineer on the United States Navy's P-3 Orion aircraft. In 1963 Brand graduated from the United States Naval Test Pilot School and was assigned to Palmdale, California as an experimental test pilot on Canadian and German F-104 programs, and prior to selection to the astronaut program he worked at the West German F-104G Flight Test Center at Istres, France as an experimental test pilot and leader of a Lockheed flight test advisory group.

He was one of the nineteen pilot astronauts selected by NASA in April of 1966, and was initially a crew member in the thermal vacuum chamber testing of the prototype command module, and a support crewman on Apollos 8 and 13. Later he was backup command module pilot for Apollo 15, and backup commander for Skylabs 3 and 4. When Skylab 3's CSM had problems with its Reaction Control System, Brand was put on standby to pilot a rescue mission, but was later stood down when it was decided the problem did not require the rescue mission to be launched. As an astronaut he also held management positions relating to spacecraft development, acquisition, flight safety and mission operations. Brand flew on four space missions - Apollo-Soyuz, STS-5, STS-41-B, and STS-35, and has logged 746 hours in space.

Brand was launched on his first space flight on July 15, 1975 as Apollo command module pilot on the Apollo-Soyuz Test Project mission. This flight resulted in the historic meeting in space between American astronauts and Soviet cosmonauts. Other crewmen on this nine-day Earth-orbital mission were Thomas Stafford, Apollo commander, Deke Slayton, Apollo docking module pilot; cosmonaut Alexei Leonov, Soyuz commander; and cosmonaut Valery Kubasov, Soyuz flight engineer. The Soyuz spacecraft was launched at Baikonur Cosmodrome, and the Apollo was launched seven hours later at the Kennedy Space Center. Two days later, the two spacecraft docked successfully. The linkup tested a unique new docking system, and demonstrated international cooperation in space. There were forty-four hours of docked joint activities which included four crew transfers between the Apollo and the Soyuz. Six records for docked and group flight were set on the mission and are recognized by the Fédération Aéronautique Internationale. Apollo splashed down in the Pacific Ocean near Hawaii on July 25, and was promptly recovered by *USS New Orleans* after completing a 217-hour mission.

Brand was commander of Space Shuttle *Columbia* for STS-5, the first fully operational flight of the Shuttle Transportation System which launched on November 11, 1982. He also commanded *Challenger* with a crew of five on the tenth flight of the Space Shuttle, STS-41-B, in 1984. Brand again commanded *Columbia* on the thirty-eighth flight of the Shuttle, this time with a crew of seven, on STS-35. At the time of the final mission Brand, at fifty-nine, became the oldest astronaut in space, later to be superseded by fellow Marines Story Musgrave, who was sixty-one, in 1996 and by John Glenn, who was seventy-seven, in 1998.

Brand departed the Astronaut Office in 1992 to become Chief of Plans at the National Aerospace Plane (NASP) Joint Program Office at Wright-Patterson Air Force Base. In September of 1994 he moved to California to become Assistant Chief of Flight Operations at the Dryden Flight Research Center, then Acting Chief Engineer, Deputy Director for Aerospace Projects, and Acting Associate Center Director for Programs. He retired from NASA in January of 2008.

RANDOLPH J. BRESNIK
Atlantis

Randolph James "Komrade" Bresnik is a Lieutenant Colonel in the United States Marine Corps and a NASA astronaut. A naval aviator by trade, Bresnik was selected as a member of NASA Astronaut Training Group 19 in May 2004. Bresnik completed his Astronaut Candidate Training in February 2006.

Bresnik was born on September 11, 1967 in Fort Knox, Kentucky, but he considers Santa Monica, California to be his hometown. He graduated from Santa Monica High School in 1985, earned a Bachelor of Arts degree in Mathematics from The Citadel in 1989, and later earned a Master of Science degree in Aviation Systems from the University of Tennessee-Knoxville in 2002. He is the first graduate of The Citadel to have the opportunity to fly in space.

In May of 1989 Bresnik received his commission as a Second Lieutenant in the Marine Corps from the Naval Reserve Officer Training Corps at The Citadel. After graduation he attended The Basic School and the Infantry Officers Course at Quantico, and following Aviation Indoctrination and primary flight training at Pensacola he entered Intermediate and Advanced flight training in Beeville, Texas and was designated a Naval Aviator in 1992.

Bresnik then reported to the Navy Fighter/Attack Training Squadron VFA-106, Naval Air Station Cecil Field, for initial

F/A-18 instruction. Upon completion of training he reported to Marine Fighter/Attack Squadron VMFA-212 at Marine Corps Air Station Kaneohe Bay, Hawaii, then MCAS El Toro and MCAS Miramar in California, where he made three overseas deployments to the Western Pacific. While assigned to VMFA-212 he attended the Marine Corps Weapons and Tactics Instructors Course (WTI) and Naval Fighter Weapons School (Top Gun).

Bresnik was selected for Naval Test Pilot School at NAS Patuxent River, Maryland and began the course in January 1999. After graduation he was assigned as an F/A-18 Test Pilot/Project Officer at the Naval Strike Aircraft Test Squadron, and while at Strike flew the F/A-18 A-D and F/A-18 E/F in all manners of flight testing.

In January of 2001 he returned to the USNTPS as a Fixed-Wing and Systems Flight Instructor, where he instructed in the F/A-18, T-38, and T-2. Bresnik returned to NSATS in January 2002 to continue flight test on the F/A-18 A-F as the Platform/Project Coordinator.

In November of 2002 he reported to Marine Aircraft Group 11 as the Future Operations Officer, and in January 2003 MAG-11 deployed to Ahmed Al Jaber Air Base, Kuwait. From Al Jaber he flew combat missions in the F/A-18 with VMFA(AW)-225 in support of Operation Southern Watch and Operation Iraqi Freedom.

Bresnik was selected by NASA in May of 2004 as an astronaut candidate, one of two pilots chosen in the Astronaut Class of 2004. In February 2006 he completed Astronaut Candidate Training, and in November 2009 he flew as a Mission Specialist on STS-129 during which he participated in the second and third spacewalks of the mission totaling 11 hours and 50 minutes. His wife Rebecca gave birth to their daughter Abigail while he was in orbit.

JAMES F. BUCHLI
Atlantis, Challenger, and Discovery

James Frederick Buchli is an engineer, retired Marine Corps aviator, and a former NASA astronaut who flew on four space shuttle missions.

Buchli was born on June 20, 1945 in New Rockford, North Dakota and received his commission in the Marine Corps following graduation from the United States Naval Academy at Annapolis, Maryland in 1967. He graduated from Basic Infantry Officer's Course and was subsequently sent to the Republic of Vietnam for a one year tour of duty, where he served as a Platoon Commander with the 9th Marine Regiment and then as Executive Officer and Company Commander for B Company, 3rd Reconnaissance Battalion.

He returned to the United States in 1969 for naval flight officer training at Naval Air Station Pensacola, Florida, and after earning his wings spent the next two years assigned to VMFA-122 at Marine Corps Air Station Kaneohe Bay, Hawaii, and Marine Corps Air Station Iwakuni, Japan. In 1973 he reported for duty with VMFA-115 at Royal Thai Air Base Nam Phong in Thailand, and then did another tour at MCAS Iwakuni. Upon completing this tour of duty he again returned to the United States and participated in the Marine Advanced Degree Program at the University of West Florida. He was subsequently assigned to VMFA-312 at Marine Corps Air Station Beaufort, South Carolina, and in 1977 attended the United States Naval Test Pilot School at Naval Air Station Patuxent River, Maryland.

Buchli became a NASA astronaut in August of 1979 after selection as part of Group 8. He was a member of the support crew for STS-1 and STS-2, and was On-Orbit CAPCOM for STS-2. A veteran of four space flights, Buchli has orbited the earth 319 times, traveling 7.74 million miles in twenty days, ten hours, twenty-five minutes and thirty-two seconds. He served as a mission specialist aboard Space Shuttle *Discovery* three times, and *Challenger* once.

On September 1, 1992 Buchli retired from the Marine Corps and the NASA Astronaut Office to accept a position as Manager, Space Station Systems Operations and Requirements with Boeing Defense and Space Group at Huntsville, Alabama. In April of 1993 he was reassigned as Boeing Deputy for Payload Operations, Space Station Freedom Program, and he currently serves as Operations & Utilization Manager for Space Station, Boeing Defense and Space Group in Houston, Texas.

James Buchli has logged over 4,200 hours flying time, with 4,000 hours in jet aircraft including combat in the F-4 Phantom II. His awards include the Defense Superior Service Medal, Legion of Merit, Purple Heart, Defense Meritorious Service Medal, Air Medal, Navy & Marine Corps Commendation Medal, NASA Distinguished Service Medal, and NASA Exceptional Service Medal.

ROBERT D. CABANA
Discovery and Columbia

Colonel Robert Donald Cabana is the Director of NASA's John F. Kennedy Space Center, a former astronaut, a veteran of four Space Shuttle flights, and a retired Marine Corps Naval Flight Officer.

Cabana was born on January 23, 1949 in Minneapolis, Minnesota to Ted and Annabell Cabana, who still reside there. In 1967 he graduated from Washburn High School in Minneapolis, and in 1971 he received a Bachelor of Science degree in mathematics from the United States Naval Academy.

After graduation from the Naval Academy Cabana attended The Basic School at Marine Corps Base Quantico, Virginia, and completed Naval Flight Officer training at Naval Air Station Pensacola, Florida in 1972. He served as an A-6 Intruder bombardier/navigator with squadrons in the 2nd Marine Aircraft Wing at Marine Corps Air Station Cherry Point, North Carolina, and the 1st Marine Aircraft Wing at Marine Corps Air Station Iwakuni, Japan. He returned to NAS Pensacola in 1975 for pilot training and was designated a naval aviator in September of 1976, at which time he received the Daughters of the American Revolution Award as the top Marine to complete naval flight training.

Cabana was then assigned to the 2nd MAW at MCAS Cherry Point, North Carolina where he flew A-6 Intruders. He later graduated from the United States Naval Test Pilot

School in 1981 as the Distinguished Graduate, and served at the Naval Air Test Center at Naval Air Station Patuxent River, Maryland as the A-6 program manager, X-29 advanced technology demonstrator project officer, and as a test pilot for flight systems and ordinance separation testing on A-6 and A-4 Skyhawk series aircraft. Prior to his selection as an astronaut candidate he was serving as the Assistant Operations Officer of Marine Aircraft Group 12 at MCAS Iwakuni, Japan.

Selected by NASA as an astronaut candidate in June 1985, Cabana completed initial astronaut training in July 1986 and qualified for assignment as a pilot on future Space Shuttle flight crews. His initial assignment was as the Astronaut Office Space Shuttle flight software coordinator until November 1986, followed by a tour as the Deputy Chief of Aircraft Operations for the Johnson Space Center where he served for two and a half years. He then served as the lead astronaut in the Shuttle Avionics Integration Laboratory (SAIL) where the Orbiter's flight software is tested prior to flight. Cabana has also served as a spacecraft communicator (CAPCOM) in Mission Control during Space Shuttle missions, and as Chief of Astronaut Appearances. Prior to his assignment to command STS-88 Cabana served three years as the Chief of NASA's Astronaut Office, and afterward he served as the Deputy Director of Flight Crew Operations. After joining the ISS Program in October 1999 Cabana was named Manager for International Operations, and from August 2001 to September 2002 he served as Director, Human Space Flight Programs, Russia. As NASA's lead representative to the Russian Aviation and Space Agency (Rosaviakosmos) and its contractors, he provided oversight of all human space flight operations, logistics, and technical functions, including NASA's mission operations in Korolev and crew training at the Gagarin Cosmonaut Training Center in Star City. Upon his return to Houston Cabana was assigned briefly as the Deputy Manager, International Space Station (ISS) Program, and from November 2000 to March 2004 served as Director, Flight Crew Operations Directorate, making him responsible for directing the day-to-day activities of the directorate, including the NASA Astronaut Corps and aircraft operations at Ellington Field. From October 2007 through October 2008 Cabana served as Director, John C. Stennis Space Center, and in October of 2008 he was reassigned as Director of the John F. Kennedy Space Center.

STS-41 *Discovery* launched in October of 1990, STS-53 *Discovery* launched from the Kennedy Space Center in December of 1992, and STS-65 *Columbia* launched in July of 1994. STS-88 *Endeavour*, Cabana's final mission, blasted off in December of 1998 and was the first International Space Station assembly mission.

Cabana retired from the Marine Corps in August of 2000 after logging over 7,000 hours in thirty-four different kinds of aircraft, and another 1,010 hours in space. His awards include the Defense Superior Service Medal, Distinguished Flying Cross, Defense Meritorious Service Medal, Meritorious Service Medal, National Intelligence Achievement Medal, NASA Distinguished Service Medal, two NASA Outstanding Leadership Medals, and two NASA Exceptional Service Medals.

KENNETH D. CAMERON
Atlantis and Discovery

Colonel Kenneth Donald Cameron is a retired Marine and NASA astronaut.

Cameron was born on November 29, 1949 in Cleveland, Ohio. He graduated from Rocky River High School in 1967 and entered the Massachusetts Institute of Technology (MIT), where he received a Bachelor of Science degree in aeronautics and astronautics in 1978, and a Master of Science degree in aeronautics and astronautics in 1979. He graduated from United States Navy Test Pilot School in 1983, and completed numerous courses in Russian language and Russian space systems at MIT, JSC, and at the Gagarin Cosmonaut Training Center in Moscow, Russia. Then in 2002 he received a Master of Business Administration degree from Michigan State University.

Cameron was commissioned in the Marine Corps in 1970 at Officer Candidate School aboard Marine Corps Base Quantico, Virginia. After graduating from the Infantry Officer's Course and Vietnamese Language School he was assigned to the Republic of Vietnam for a one-year tour of duty as an infantry platoon commander with the 1st Battalion, 5th Marines, and later with the Marine Security Guards at the U.S. Embassy in Saigon. Upon his return to the United States he served as Executive Officer with India Company, 3rd Battalion, 2nd Marines at Camp Lejeune,

North Carolina. He reported to Naval Air Station Pensacola, Florida in 1972 for flight training, and received his Naval Aviator wings in 1973. He was then assigned to VMA-223 at Marine Corps Air Station Yuma, Arizona flying A-4M Skyhawks.

In 1976 Cameron was reassigned to the Massachusetts Institute of Technology, where he participated in the Marine College Degree and Advanced Degree Programs. Upon graduation he was assigned to flying duty for one year with Marine Aircraft Group 12 at Marine Corps Air Station Iwakuni, Japan. He was subsequently assigned to the Pacific Missile Test Center in 1980, and in 1982 to the United States Naval Test Pilot School, Naval Air Station Patuxent River, Maryland. Following graduation in 1983 he was assigned as project officer and test pilot in the F/A-18 Hornet, A-4, and OV-10 Bronco airplanes with the Systems Engineering Test Directorate at the Naval Air Test Center.

Selected by NASA in May of 1984, Cameron became an astronaut in June of 1985. His technical assignments have included work on Tethered Satellite Payload, flight software testing in the Shuttle Avionics Integration Laboratory (SAIL), launch support activities at Kennedy Space Center, and spacecraft communicator (CAPCOM) in Mission Control. His management assignments in NASA include Section Chief for astronaut software testing in SAIL, astronaut launch support activities, and Operations Assistant to the Hubble Repair Mission Director. In 1994 Cameron served as the first NASA Director of Operations in Star City, Moscow, where he worked with the Cosmonaut Training Center staff to set up a support system for astronaut operations and training in Star City, and received Russian training in Soyuz and Mir spacecraft systems and flight training in Russian aircraft.

Cameron flew his first mission as a shuttle pilot on STS-37, launched in April of 1991. On his second mission he was spacecraft commander on STS-56, which launched in April of 1993 while carrying ATLAS-2. During this nine-day mission the crew of *Discovery* conducted atmospheric and solar studies in order to better understand the effect of solar activity on the Earth's climate and environment, and deployed and retrieved the autonomous observatory *Spartan*. On his third mission Cameron commanded *Atlantis* on STS-74, which launched in November of 1995. It was NASA's second space shuttle mission to rendezvous and dock with the Russian Space Station Mir, and the first mission to use the shuttle to assemble a module and attach it to a space station.

Cameron left NASA on August 5, 1996 to join Hughes Training, Inc., a subsidiary of General Motors Corporation, as Executive Director of Houston Operations. In September of 1997 he transferred to Saab Automobile in Sweden as Vehicle Line Executive for the Saab 9-3 automobile, and upon his return to the U.S. worked at the GM Technical Center near Detroit in positions in Purchasing and Research and Development.

Cameron returned to the space program in October of 2003, taking a position as Principal Engineer in the NASA Engineering & Safety Center, based at the NASA Langley Research Center, in Hampton, Virginia.

Colonel Cameron has logged over 4,000 hours flying time in forty-eight different types of aircraft, and has logged over 561 hours in space. His decorations include the Legion of Merit, Defense Superior Service Medal, two Distinguished Flying Crosses, Navy Commendation Medal with Combat "V," NASA Leadership Medal, and NASA Exceptional Achievement Medal.

GERALD P. CARR
Apollo and Skylab

Gerald Paul Carr is an engineer, retired Marine Corps Colonel and former NASA astronaut.

Carr was born in Denver, Colorado on August 22, 1932 but raised in Santa Ana, California, which he considers his hometown. He graduated from Santa Ana High School, and received a Bachelor of Engineering degree in Mechanical Engineering from the University of Southern California in 1954, a Bachelor of Science in Aeronautical Engineering from the U.S. Naval Postgraduate School in 1961, and a Master of Science degree in Aeronautical Engineering from Princeton University in 1962. He was also presented an Honorary Doctorate of Science, Aeronautical Engineering, from Parks College of Saint Louis University in Cahokia, Illinois in 1976.

Carr began his military career in 1949 with the Navy, and in 1950 was appointed a midshipman with the Naval Reserve Officer Training Corps detachment at the University of Southern California. Upon graduation in 1954 he received his commission, and subsequently reported to The Basic School at Marine Corps Base Quantico, Virginia. He received flight training at Naval Air Station Pensacola, Florida, and Naval Air Station Kingsville, Texas, and was then assigned to VMF(AW)-114 where he gained experience in the F9F Cougar and the F-6A Skyray. After postgraduate training, he served with VMFA(AW)-122 from 1962 to

1965, piloting the F-8 Crusader in the United States and Far East.

Carr was one of the nineteen astronauts selected by NASA in April of 1966. When he was informed by NASA of his selection for astronaut training he was assigned to the test directors section of Marine Air Control Squadron Three, a unit responsible for the testing and evaluation of the Marine Tactical Data System. He served as a member of the astronaut support crews and as CAPCOM for the Apollo 8 and 12 flights, and was involved in the development and testing of the lunar roving vehicle which was used on the lunar surface by Apollo flight crews. He was in the likely crew rotation to serve as Lunar Module Pilot for Apollo 19, before the program was cancelled by NASA in 1970.

Carr was commander of *Skylab 4*, which was the third and final manned visit to the Skylab Orbital Workshop, between November of 1973 and February of 1974. Carr and his *Skylab 4* teammates shared the world record for individual time in space, 2,017 hours 15 minutes 32 seconds, and Carr himself logged 15 hours and 48 minutes in three EVAs outside the Orbital Workshop. He was accompanied on the record-setting 34.5-million-mile flight by Dr. Edward Gibson and William R. Pogue, and together they logged 338 hours of operation of the Apollo Telescope Mount, which made extensive observations of the sun's solar processes.

In mid-1977 Carr was named head of the design support group within the astronaut office, and was responsible for providing crew support to such activities as space transportation system design, simulations, testing, and safety assessment, and for development of man/machine interface requirements.

Carr retired from the Marine Corps in 1975 and from NASA in 1977, and founded CAMUS Incorporated in 1984.

The family-owned corporation provides technical support services in Zero-G Human Factors Engineering, Procedures Development, Operations Analysis, and Training and Systems Integration. CAMUS was a major contributor as a technical support subcontractor to the Boeing Company in the crew systems design of the International Space Station. In addition, the corporation is involved in the production of fine art designed by his wife, artist/sculptor Pat Musick.

Colonel Carr has logged more than 8,000 flying hours, 5,365 hours of which are jet time, and was inducted into the Astronaut Hall of Fame in 1997. He is also a Fellow of the American Astronautical Society, former Director of the Sunsat Energy Council, former Director of the Houston Pops Orchestra, and a Director of the National Space Society.

In 1975 the Gerald P. Carr Intermediate School (previously Ralph C. Smedley Junior High) in Santa Ana, California was renamed in his honor, and the school's team name was changed to the Astros in honor of Carr's NASA achievements.

R. WALTER CUNNINGHAM
Apollo

Colonel Ronnie Walter "Quincy" Cunningham is a retired Marine fighter pilot and astronaut, as well as a physicist, lecturer, venture capitalist, author and host of the radio talk show *Lift-off To Logic.* He was NASA's second civilian astronaut, and in 1968 he was the Lunar Module pilot for the *Apollo 7* mission.

Cunningham was born in Creston, Iowa on March 16, 1932 and graduated from Venice High School - where he now has a building named for him - in California. After high school he joined the U.S. Navy in 1951, and began flight training in 1952. He then served on active duty as a fighter pilot with the Marine Corps from 1953 until 1956, and from 1956 to 1975 he served in the Marine Corps Reserve program and eventually retired at the rank of Colonel.

Cunningham received his Bachelor of Arts and Literature degree in 1960, and his Master of Arts degree in 1961, both in physics, from the University of California at Los Angeles. He then worked as a scientist for the Rand Corporation.

In October of 1963 Cunningham was one of the third group of astronauts selected by NASA. At the time he was NASA's second civilian astronaut, and on October 11, 1968 he occupied the lunar module pilot seat for the eleven-day flight of *Apollo 7*. Although the flight carried no lunar module, Cunningham was kept busy with the myriad system tests aboard this first launch of a manned Apollo mission. He left NASA in 1971, graduated from Harvard Business School's Advanced Management Program in 1974, and worked as a businessman and investor in a number of private ventures.

In 1977 Cunningham published *The All-American Boys*, a reminiscence of his astronaut days. He was also a major contributor and foreword-writer for the 2007 space history book *In the Shadow of the Moon*.

Cunningham is currently a sometimes controversial radio personality and public speaker, and is well known as a vocal global warming skeptic. He was depicted on film in the 1998 miniseries *From the Earth to the Moon*, with his character being played by Fredric Lehne.

In 2008 NASA corrected a longtime oversight and retroactively awarded Cunningham the NASA Distinguished Service Medal for his part in the *Apollo 7* mission.

FRED HAISE
Apollo and Enterprise

Fred Wallace Haise, Jr. is an engineer, former astronaut, and one of only twenty-four people to have flown to the Moon.

Haise was born in Biloxi, Mississippi, where he attended Biloxi High School and Perkinston Junior College (now Mississippi Gulf Coast Community College). He also graduated with honors in aeronautical engineering from the University of Oklahoma in 1959, completed post-graduate courses at the USAF Aerospace Test Pilot School at Edwards Air Force Base in 1964, and the Harvard Business School PMD Program in 1972.

Haise completed naval aviator training in 1954 and served as a United States Marine Corps fighter pilot. His NASA career began as an aeronautical research pilot at Lewis Research Center in 1959, and further assignments included serving as a research pilot at the NASA Dryden Flight Research Center in 1963, and as an astronaut at the Johnson Space Center in 1966. Haise was the first of the 1966 group to be assigned to Apollo duties, and was placed ahead of some group 3 members. He served on the back-up crews for the *Apollo 8*, *Apollo 11*, and *Apollo 16* moon missions.

Haise flew as the lunar module pilot on the aborted *Apollo 13* lunar mission in 1970. Due to the free return trajectory on this mission Haise, along with Jim Lovell and Jack Swigert, the other two astronauts on *Apollo 13*, most likely hold the record for the furthest distance ever traveled from the earth by human beings.

Haise was also scheduled to be commander of the cancelled *Apollo 19* mission. He later flew five flights as the commander of the space shuttle *Enterprise* in 1977 for the Approach and Landing Tests Program at Edwards Air Force Base, and was selected to command the original STS-2 mission to rescue the *Skylab* space station in 1979 - but it was cancelled due to long delays in the Shuttle's development and the break-up of the *Skylab* in mid-1979.

Fred Haise was presented with the Presidential Medal of Freedom in 1970 by President Richard Nixon, and retired from NASA in June of 1979 to become a manager with Grumman Aerospace before retiring in 1996. In 1995 Fred Haise was inducted into the Aerospace Walk of Honor.

Bill Paxton played the role of Haise in the 1995 film *Apollo 13*, and Adam Baldwin played him in the mini-series *From The Earth To The Moon.*

DAVID C. HILMERS
Atlantis and Discovery

David Carl Hilmers is a former Maine Corps pilot and NASA astronaut.

Hilmers was born on January 28, 1950 in Clinton, Iowa and graduated from Central Community High School in DeWitt, Iowa in 1968. He later received a Bachelor of Arts degree in mathematics (Summa Cum Laude) from Cornell College in 1972, a Master of Science degree in electrical engineering (with distinction) in 1977, and an electrical engineering degree from the U.S. Naval Postgraduate School in 1978. He also has received an MD degree from the Baylor College of Medicine in 1995, and a Master of Science degree from the University of Texas Houston Health Science Center in 2002.

Hilmers entered active duty with the Marine Corps in July of 1972, and upon completing The Basic School and Naval Flight Officer School was assigned to VMA(AW)-121 at Marine Corps Air Station Cherry Point, North Carolina flying the A-6 Intruder as a bombardier-navigator. In 1975 he became an air liaison officer with the 1st Battalion, 2nd Marines, which was stationed with the 6th Fleet in the Mediterranean. He graduated from the Naval Postgraduate School in 1978, was later assigned to the 1st Marine Aircraft Wing at Marine Corps Air Station Iwakuni, Japan, and was stationed with the 3rd Marine Aircraft Wing at Marine Corps

Air Station El Toro, California at the time of his selection by NASA.

Hilmers was selected a NASA astronaut in July of 1980, completed the initial training period in August of 1981, and in 1983 was selected as a member of the launch-ready standby crew. His early NASA assignments have included work on upper stages such as PAM, IUS, and Centaur, as well as shuttle software verification at the Shuttle Avionics Integration Laboratory (SAIL). In addition, he was the Astronaut Office training coordinator, worked on various Department of Defense payloads, served as a spacecraft communicator (CAPCOM) at Mission Control for STS-41D, STS-41G, STS-51A, STS-51C and STS-51D, worked Space Station issues for the Astronaut Office, and served as head of the Mission Development Branch within the Astronaut Office.

In May of 1985 he was named to the crew of STS-61F, which was to deploy the Ulysses spacecraft on an interplanetary trajectory using a Centaur upper stage. This mission was to have flown in May of 1986, but the Shuttle Centaur project was terminated in July 1986 and Hilmers then worked in the areas of ascent abort development, payload safety, and shuttle on-board software. During 1987 he was involved in training for STS-26 and in-flight software development.

His missions include STS-51-J *Atlantis*, a classified Department of Defense mission launched in October of 1985, STS-26 *Discovery*, the first mission to be flown after the *Challenger* accident, in 1988, STS-36 *Atlantis* in February of 1990, and STS-42 *Discovery* in 1992.

David Hilmers retired from NASA in 1992 and is now a faculty member of the Medicine/Pediatrics Department of the Baylor College of Medicine in Houston, Texas.

CHARLES O. HOBAUGH
Atlantis and Endeavour

Charles Owen "Scorch" Hobaugh is a Marine Corps officer and NASA astronaut.

Hobaugh was born on November 5, 1961 in Bar Harbor, Maine and graduated from North Ridgeville High School in North Ridgeville, Ohio in 1980. In 1984 he received a Bachelor of Science degree in Aerospace Engineering from the U.S. Naval Academy, and in 1994 graduated from the University of Tennessee Space Institute.

Hobaugh received his commission as a Second Lieutenant in the Marine Corps at the United States Naval Academy in May of 1984. He graduated from The Basic School in December of 1984, and after a six month temporary assignment at the Naval Air Systems Command reported to Naval Aviation Training Command and was designated a Naval Aviator in February of 1987. He then reported to VMAT-203 for initial AV-8B Harrier training, and upon completion of this training was assigned to VMA-331. He then made overseas deployments to Marine Corps Air Station Iwakuni, Japan and flew combat missions in the Persian Gulf during Operation Desert Storm while embarked aboard *USS Nassau*. While assigned to VMA-331 he also attended the Marine Aviation Warfare and Tactics Instructor Course and was subsequently assigned as the Squadron Weapons and Tactics Instructor.

Hobaugh was selected for U.S. Naval Test Pilot School and began the course in June of 1991. After graduation in June of 1992 he was assigned to the Strike Aircraft Test Directorate as an AV-8 Project Officer and as the ASTOVL/JAST/JSF Program Officer. While there he flew the AV-8B, YAV-8B (VSRA) and A-7E. Then in 1994 he went back to the Naval Test Pilot School as an instructor in the Systems Department where he flew the F-18, T-2 Buckeye, U-6A and gliders, and while there he was selected for the astronaut program.

Selected by NASA in April 1996, Hobaugh reported to the Johnson Space Center in August 1996. He completed two years of training and evaluation and was qualified for flight assignment as a pilot, but was initially assigned technical duties as part of the Astronaut Office Spacecraft Systems/Operations Branch. Projects included Landing and Rollout, evaluator in the Shuttle Avionics Integration Laboratory, Advanced Projects, Multifunction Electronics Display Enhancements, Advanced and Upgrade, Rendezvous and Close Proximity Operations and Visiting Vehicles prior to his first flight assignment. Most recently he served as Capsule Communicator, working in Mission Control as the voice to the crew.

Hobaugh was the reentry and landing CAPCOM for the STS-107 mission on which the Space Shuttle *Columbia* was destroyed on reentry, and spoke the words, "Columbia, Houston. UHF Comm Check" several times after Mission Control had lost contact with *Columbia*.

Hobaugh flew as the pilot of STS-104 in July of 2001, the tenth mission to the International Space Station. During the thirteen-day flight the crew conducted joint operations with the Expedition 2 crew, and performed three spacewalks to install the Quest Joint Airlock and outfit it with four high-pressure gas tanks. The mission was accomplished in two hundred Earth orbits and traveled 5.3 million miles in 306 hours and 35 minutes.

Hobaugh later flew as pilot on STS-118 in August of 2007 for thirteen days, and served as commander of the STS-129 mission aboard Space Shuttle *Atlantis* for ten days in November 2009.

Prior to the launch of STS-118 several reporters asked Hobaugh the meaning of his call-sign, "Scorch." He has yet to explain it, and says he probably won't because he feels it sounds better if you don't know the origin of the name – although he did admit it is related to his days of flying Harrier jets in the Marine Corps.

Hobaugh has logged over 3,000 flight hours in more than forty different aircraft, and has over 200 V/STOL shipboard landings. He is also a U.S. Naval Academy Distinguished Graduate, U.S. Naval Test Pilot School Distinguished Graduate, and has received the Joe Foss Award for Advanced Jet Training.

DOUGLAS G. HURLEY
Endeavour

Douglas Gerald Hurley is a Marine Corps aviator and NASA astronaut. He piloted Space Shuttle mission STS-127 in 2009, and is the first Marine to fly the F/A-18 E/F Super Hornet.

Hurley was born on October 21, 1966 in Endicott, New York and graduated from Owego Free Academy in Owego, New York in 1984. He later graduated Magna Cum Laude with Honors from Tulane University, earning his B.S.E. in Civil Engineering in 1988, and was also a Distinguished Graduate from Tulane's NROTC program.

Hurley was a Distinguished Graduate from Officer Candidate School, and received his commission as a Second Lieutenant in the Marine Corps from the Naval Reserve Officer Training Corps at Tulane University in 1988. After graduation he attended The Basic School and later the Infantry Officers Course at Marine Corps Base Quantico, Virginia. Following Aviation Indoctrination at Naval Air Station Pensacola, Florida, he entered flight training in Texas in 1989 and was a Distinguished Graduate of the U.S. Navy Pilot Training program. He was designated a Naval Aviator in August of 1991.

Hurley then reported to VMFAT-101 at Marine Corps Air Station El Toro, California for initial F/A-18 Hornet training, after which he was assigned to VMFA(AW)-225 and made three overseas deployments to the Western Pacific. While assigned to VMFA(AW)-225 he attended the United States Marine Aviation Weapons and Tactics Instructor Course, the Marine Division Tactics Course, and the Aviation Safety Officers Course at the Naval Postgraduate School in Monterey, California. Over his four and a half years with the "Vikings," he served as the Aviation Safety Officer and the Pilot Training Officer.

Hurley was then selected to attend the United States Naval Test Pilot School at Naval Air Station Patuxent River, Maryland and began the course in January of 1997. After graduation in December of that year he was assigned to the Naval Strike Aircraft Test Squadron (VX-23) as an F/A-18 Project Officer and Test Pilot. At "Strike," he participated in a variety of flight testing including flying qualities, ordnance separation, and systems testing and became the first ever Marine pilot to fly the F/A-18 E/F Super Hornet. He was serving as the Operations Officer when selected for the astronaut program.

After selection as a pilot by NASA in July of 2000 Lieutenant Colonel Hurley reported for training in August of that year. Following the completion of two years of training and evaluation, he was assigned technical duties in the Astronaut Office which have included Kennedy Operations Support as a "Cape Crusader" where he was the lead ASP (Astronaut Support Personnel) for Shuttle missions STS-107 and STS-121. He has also worked Shuttle Landing and Rollout, served on the Columbia Reconstruction Team at Kennedy Space Center, and in the Exploration Branch in support of the selection of the Orion Crew Exploration Vehicle. More recently he served as the NASA Director of Operations at the Gagarin Cosmonaut Training Center in Star City, Russia.

In July of 2009 Hurley served as the pilot of STS-127 aboard *Endeavour* for ISS Assembly Mission 2J/A, which delivered the Japanese-built Exposed Facility and the Experiment Logistics Module Exposed Section to the International Space Station.

Hurley has logged over 3200 hours in more than twenty-two aircraft, and is a recipient of the Stephen A. Hazelrigg Memorial Award for best Test Pilot/Engineer Team, Naval Strike Aircraft Test Squadron. He has been awarded the Meritorious Service Medal, two Navy and Marine Corps Commendation Medals, and various other service awards.

JACK R. LOUSMA
Skylab and Columbia

Jack Robert Lousma is a former Marine Corps pilot and NASA astronaut who was a member of the second manned crew on the *Skylab* space station and also commanded the third space shuttle mission.

Lousma was born on February 29, 1936 in Grand Rapids, Michigan and graduated from Ann Arbor High School. He earned a Bachelor of Science degree in Aeronautical Engineering from the University of Michigan in 1959, and an M.S. degree in Aeronautical Engineering from the U. S. Naval Postgraduate School in 1965. He had also been presented with an honorary Doctorate of Astronautical Science from the University of Michigan in 1973, an honorary D.Sc. from Hope College in 1982, and an honorary D.Sc. in Business Administration from Cleary College in 1986.

Lousma was a reconnaissance pilot with VMCJ-2 of the 2nd Marine Aircraft Wing at Marine Corps Air Station Cherry Point, North Carolina before being assigned to Houston and the Lyndon B. Johnson Space Center. He became a Marine Corps officer in 1959, and received his wings in 1960 after completing training at the U.S. Naval Air Training Command. He was then assigned to VMA-224, 2nd MAW as an attack pilot, and later served with VMA-224 of the 1st Marine Air Wing at Marine Corps Air Station Iwakuni, Japan.

Lousma was one of the 19 astronauts selected by NASA in April 1966, and served as a member of the astronaut support crews for the *Apollo 9, 10,* and *13* missions. He famously was the CAPCOM recipient of the "Houston, we have a problem" message from *Apollo 13.* He was the pilot for *Skylab 3* from July 28 to September 25, 1973, and was commander of *Columbia* for STS-3 in 1982, logging a total of over 1,619 hours in space and eleven hours on two spacewalks outside the *Skylab* space station. He also served as backup docking module pilot of the United States flight crew for the Apollo-Soyuz Test Project mission which was completed successfully in July of 1975..

The crew for the 59-1/2 day flight of *Skylab 3* included spacecraft commander Alan L. Bean, pilot Lousma, and science-pilot Owen K. Garriott. SL-3 accomplished 100% of its mission goals while completing 858 revolutions of the earth and traveling some 24,400,000 miles in earth orbit. The crew installed six replacement rate gyros used for attitude control of the spacecraft, a twin-pole sunshade used for thermal control, and repaired nine major experiment or operational equipment items. They devoted 305 man hours to extensive solar observations from above the Earth's atmosphere, which included viewing two major solar flares and numerous smaller flares and coronal transients. Also acquired and returned to earth were 16,000 photographs and eighteen miles of magnetic tape documenting earth resource observations. The crew completed 333 medical experiment performances, and obtained valuable data on the effects of extended weightlessness on humans. *Skylab-3* ended with a Pacific Ocean splashdown and recovery by *USS New Orleans.*

STS-3, the third orbital test flight of space shuttle *Columbia*, launched on March 22, 1982 into a 180-mile circular orbit above the earth. Lousma was the spacecraft commander, and C. Gordon Fullerton was the pilot, on this eight-day mission. Major flight test objectives included exposing *Columbia* to extremes in thermal stress and the first use of the fifty-foot Remote Manipulator System to grapple and maneuver a payload in space. The crew also operated several scientific experiments in the orbiter's cabin and on the OSS-1 pallet in the payload bay. Space Shuttle *Columbia* responded favorably to the thermal tests, and was found to be better than expected as a scientific platform. The crew accomplished almost one hundred percent of the objectives assigned to STS-3 while traveling 3.4 million miles during 129.9 orbits of the earth. Mission duration was 192 hours, 4 minutes, and 49 seconds, and after a one-day delay due to bad weather *Columbia* landed on the lakebed at White Sands, New Mexico on March 30, 1982 and became the only shuttle to ever land there.

Lousma left NASA in 1983, and in 1984 ran for the U.S. Senate as a Republican against Carl Levin, the incumbent Senator from Michigan, but lost after receiving 47% of the vote. Lousma survived a bitter primary fight against former Republican Congressman Jim Dunn to capture the nomination, and although Ronald Reagan's landslide reelection was a boon to his chances he was hurt late in the campaign when video surfaced of him telling a group of Japanese auto manufactures that his family owned a Japanese-made car. This did not play well in the Detroit area.

In 1988 Lousma commented on the STS-26 launch for ITN on British television, reflecting the media interest in the first Shuttle flight following the *Challenger* accident. During the ascent, as Lousma described the abort modes as they became available, the show's host Alastair Burnet quickly asked Lousma which abort mode he preferred, and "abort to orbit" came the quick reply.

Lousma logged 7000 hours of flight time, including 700 hours in general aviation aircraft, 1619 hours in space, 4,500 hours in jet aircraft, and 240 hours in helicopters, and was portrayed by Quinn Redeker in the 1974 TV movie *Houston, We've Got a Problem.*

Jack Lousma is a fellow of the American Astronautical Society, and a member of the University of Michigan "M" Club, the Officer's Christian Fellowship and the Association of Space Explorers. He has been awarded the Navy Distinguished Service Medal, City of Chicago Gold Medal, Robert J. Collier Trophy, Marine Corps Aviation Association's Exceptional Achievement Award, NASA Distinguished Service Medal, Department of Defense Distinguished Service Medal, and has been inducted into the Michigan Aviation Hall of Fame.

FRANKLIN STORY MUSGRAVE
Challenger, Discovery, Atlantis, Endeavour, Columbia

Franklin Story Musgrave is a doctor, former Marine, and retired NASA astronaut. He is currently a public speaker and consultant to both Disney's Imagineering Group and Applied Minds in California.

Musgrave was born on August 19, 1935 in Boston, Massachusetts but considers Lexington, Kentucky to be his hometown. He attended Dexter School in Brookline, Massachusetts and St. Mark's School in Southborough from 1947 to 1953, but left school shortly before graduation and before receiving his high school diploma. He has since received a BS degree in mathematics and statistics from Syracuse University in 1958, an MBA degree in operations analysis and computer programming from the University of California, Los Angeles in 1959, a BA degree in chemistry from Marietta College in 1960, an M.D. degree from Columbia University College of Physicians and Surgeons in 1964, an MS in physiology and biophysics from the University of Kentucky in 1966, and a MA in literature from the University of Houston–Clear Lake in 1987.

Musgrave entered the Marine Corps after leaving school in 1953 and served as an aviation electrician and instrument technician and later as an aircraft crew chief while completing duty assignments in Korea, Japan, and Hawaii, as well as aboard the carrier *USS Wasp* in the Far East.

He served a surgical internship at the University of Kentucky Medical Center in Lexington from 1964 to 1965, and continued there as a U. S. Air Force post-doctoral fellow from 1965 to 1966 working in aerospace medicine and physiology, and as a National Heart Institute post-doctoral fellow between 1966 and 1967, teaching and doing research in cardiovascular and exercise physiology. From 1967 to 1989 he continued clinical and scientific training as a part-time surgeon at Denver General Hospital (presently known as Denver Health Medical Center) and as a part-time professor of physiology and biophysics at the University of Kentucky Medical Center. He has written twenty five scientific papers in the areas of aerospace medicine and physiology, temperature regulation, exercise physiology, and clinical surgery.

Musgrave was selected as a scientist-astronaut by NASA in August of 1967, completed astronaut academic training, and then worked on the design and development of the Skylab Program. He was the backup science-pilot for the first Skylab mission, and was a CAPCOM for the second and third. Musgrave participated in the design and development of all Space Shuttle extravehicular activity equipment including spacesuits, life support systems, airlocks and manned maneuvering units, and from 1979 to 1982 and again from 1983 to 1984 he was assigned as a test and verification pilot in the Shuttle Avionics Integration Laboratory at JSC.

Musgrave served as spacecraft communicator (CAPCOM) for *Challenger, Discovery, Atlantis, Endeavour, Columbia,* and lead CAPCOM for a number of subsequent flights. He was a mission specialist on STS-6 in 1983, STS-51-F/Spacelab-2 in 1985, STS-33 in 1989 and STS-44 in 1991, was the payload commander on STS-61 in 1993, and a mission specialist on STS-80 in 1996.

Musgrave is the only astronaut to have flown missions on all five Space Shuttles, and the last of the Apollo era astronauts on active flight status to retire. Prior to fellow Marine John Glenn's return to space in 1998 Musgrave held the record for the oldest person in orbit, at age sixty-two. He retired from NASA in 1997.

He has flown 17,700 hours in 160 different types of civilian and military aircraft, including 7,500 hours in jets, and has spent a total of 1281 hours 59 minutes, 22 seconds in space. He has earned FAA ratings for instructor, instrument instructor, glider instructor, and airline transport pilot in addition to his U.S. Air Force Wings. An accomplished parachutist, he has made more than eight hundred free falls, including over one hundred experimental free-fall descents involved with the study of human aerodynamics.

On top of all of these accomplishments Story Musgrave was also the subject of a bit of pop culture notoriety, when in the early 1990s he was stalked by Margaret Mary Ray, a schizophrenic woman whom had previously served time for stalking comedian David Letterman.

CARLOS I. NORIEGA
Atlantis and Endeavour

Carlos Ismael Noriega is a Peruvian born former NASA astronaut and retired Marine Corps Lieutenant Colonel.

Noriega was born on October 8, 1959 in Lima, Peru, although he considers Santa Clara, California to be his hometown. He graduated from Wilcox High School in Santa Clara in 1977, and earned a Bachelor of Science degree in computer science from the University of Southern California 1981, and Master of Science degrees in computer science and space systems operations from the Naval Postgraduate School in 1990.

Noriega was a member of the Navy ROTC unit at the University of Southern California and received his commission in the Marine Corps there in 1981. Following graduation from flight school he flew CH-46 Sea Knight helicopters with HMM-165 from 1983 to 1985 at Marine Corps Air Station Kaneohe Bay, Hawaii, and made two six-month shipboard deployments to the West Pacific/Indian Ocean including operations in support of the Multi-National Peacekeeping Force in Beirut, Lebanon. He completed his tour in Hawaii as the Base Operations Officer for Marine Air Base Squadron 24, and in 1986 was transferred to MCAS Tustin, California where he served as the aviation safety officer and instructor pilot with HMT-301. In 1988 Noriega was selected to attend the Naval Postgraduate School in Monterey, California where he earned two Master of Science degrees, and upon graduation in September of 1990 he was assigned to United States Space Command in Colorado Springs, Colorado. In addition to serving as a Space Surveillance Center Commander he was responsible for several software development projects, and was ultimately the command representative for the development and integration of the major space and missile warning computer system upgrades for Cheyenne Mountain Air Force Base. At the time of his selection, he was serving on the staff of the 1st Marine Aircraft Wing in Okinawa, Japan.

Selected by NASA in December of 1994, Noriega reported to the Johnson Space Center in March 1995. He completed a year of training and evaluation, was qualified for assignment as a mission specialist in May of 1996, and held technical assignments in the Astronaut Office EVA/Robotics and Operations Planning Branches.

Noriega flew on STS-84 in 1997, and STS-97 in 2000. Following STS-97 he trained as the backup commander for ISS Expedition 6, and later as a member of the crew of STS-121. In July of 2004 Noriega was replaced by Piers Sellers on the crew of STS-121 due to a temporary medical condition, and while awaiting future flight assignment served as Chief, Exploration Systems Engineering Division, Engineering Directorate, Johnson Space Center. In January of 2005 Noriega retired from the NASA Astronaut Corps, but he continues to serves as the Manager, Advanced Projects Office of the Constellation Program.

STS-84 was NASA's sixth Space Shuttle mission to rendezvous and dock with the Russian Space Station Mir, and during this nine-day mission the crew aboard Space Shuttle *Atlantis* conducted a number of secondary experiments and transferred nearly four tons of supplies and experiment equipment between *Atlantis* and the Mir station while logging a total of 221 hours and 20 minutes in space and traveling 3.6 million miles in 144 orbits of the Earth.

STS-97 *Endeavour*, launched in November of 2000, was the fifth space shuttle mission dedicated to the assembly of the International Space Station. While docked with the Station the crew installed the first set of U.S. solar arrays and performed three space walks, in addition to delivering supplies and equipment to the station's first resident crew.

Overall Carlos Noriega logged approximately 2,200 flight hours in various fixed-wing and rotary-wing aircraft and over 461 hours in space, including nineteen EVA hours in three space walks.

BRYAN D. O'CONNOR
Atlantis and Columbia

Colonel Bryan Daniel O'Connor is a retired Marine Corps officer and former NASA astronaut.

O'Connor was born on September 6, 1946 in Orange, California and today considers Twentynine Palms, California to be his hometown. He graduated from Twentynine Palms High School in 1964, received a Bachelor of Science degree in Engineering (minor in Aeronautical Engineering) from the United States Naval Academy in 1968, and a Master of Science degree in Aeronautical Systems from the University of West Florida in 1970. He then graduated from the Naval Safety School at the Naval Postgraduate School in Monterey, California in 1972, and from the Naval Test Pilot School in Patuxent River, Maryland in 1976.

O'Connor began active duty with the Marine Corps in June 1968 following graduation from the U.S. Naval Academy at Annapolis. He received his Naval Aviator's wings in June 1970, and served as an attack pilot flying the A-4 Skyhawk and the AV-8A Harrier on land and sea assignments in the United States, Europe and the Western Pacific.

O'Connor attended the Naval Test Pilot School in 1975 and served as a test pilot with the Naval Air Test Center's Strike Test Directorate at Patuxent River, Maryland. During this three and a half year assignment he participated in evaluations of various conventional and VSTOL aircraft, including the A-4, OV-10, AV-8, and X-22 VSTOL research aircraft. From June 1977 to June 1979 he was the Naval Air Test Center project officer in charge of all Harrier flight testing, including the planning and execution of the First Navy Preliminary Evaluation of the YAV-8B advanced Harrier prototype. When informed of his selection to NASA's Astronaut Program in 1980, he was serving as the Deputy Program Manager for the AV-8 program at the Naval Air Systems Command in Washington, D.C.

O'Connor was selected as an astronaut in May of 1980, and after a one-year initial training program at NASA's Johnson Space Center in Houston, Texas served in a variety of functions in support of the first test flights of the Space Shuttle including simulator test pilot for STS-1 and 2, safety/photo chase pilot for STS-3, support crew for STS-4, CAPCOM for STS-5 through STS-9. He also served as Aviation Safety Officer for the Astronaut Corps.

When the *Challenger* and its crew were lost in January of 1986 O'Connor was given a number of safety and management assignments over the next three years as the Space Agency recovered from the disaster. In the first days after the accident he organized the initial wreckage reassembly activities at Cape Canaveral, and then established and managed the operation of the NASA Headquarters Action Center, which is the link between NASA and the Presidential Blue Ribbon Accident Investigation Panel. In March of 1986 he was assigned duties as Assistant (Operations) to the Space Shuttle Program Manager, as well as First Chairman of NASA's new Space Flight Safety Panel, jobs he held until February 1988 and 1989 respectively. He subsequently served as Deputy Director of

Flight Crew Operations from February 1988 until August 1991.

O'Connor was the pilot on STS-61-B in 1985 and crew commander on STS-40 in 1991. STS-61-B *Atlantis* launched in November of 1985, the twenty-second shuttle flight and second-ever night shuttle launch from the Kennedy Space Center. It had the heaviest payload weight carried to orbit by the space shuttle to date, and was the first flight to deploy four satellites. STS-40 *Columbia* in June of 1991 was the first space shuttle mission dedicated to life science studies.

O'Connor left NASA in August of 1991 to become commanding officer of the Marine Aviation Detachment, Naval Air Test Center, Patuxent River. During this ten month assignment he led 110 Marine test pilots and technicians, participated as an AV-8B project test pilot, instructed students at the Test Pilot School, directed the Naval Air Test Center Museum, and became the first Marine to serve as Deputy Director and Chief of Staff of the Flight Test and Engineering Group.

O'Connor returned to NASA Headquarters in Washington after retiring from the Marine Corps to become the Deputy Associate Administrator for Space Flight. He was immediately assigned the task of developing a comprehensive flight safety improvement plan for the space shuttle, and worked closely with Congress and the Administration for funding of the major upgrade program. Then in late summer 1992 he was assigned as leader of the negotiating team that traveled to Moscow to establish the framework for what subsequently became the ambitious and complex joint manned space program known as Shuttle/MIR.

In March of 1993 O'Connor was assigned as Director, Space Station Redesign. He and his fifty-person team of engineers, managers, and international partners developed and then recommended substantial vehicle and program restructure strategies which amounted to $300 million in savings per year, thus helping to save the program from cancellation by Congress. In September he was named Acting Space Station Program Director, and he held that position throughout the transition from the Freedom Program to the new International Space Station Program and the announcement of a permanent Program Director in January of 1994.

In April 1994 O'Connor was reassigned as the Director of Space Shuttle Program, and as such was responsible for all aspects of the $3.5 billion per year program, includingd leading over 27,000 government and contractor personnel. By the time he left NASA in March of 1996 he had directed NASA's largest and most visible program through twelve safe, successful missions, including the first three flights to the Russian Space Station, planned and led an extensive program restructure designed to save the taxpayers approximately $1 billion over the five-year budget horizon, and oversaw the introduction of several major safety improvements developed to prevent another "Challenger" disaster.

O'Connor left NASA once again in February of 1996 to become an aerospace consultant. Until 2002, when he rejoined NASA as Chief Safety and Mission Assurance Officer, O'Connor served as Director of Engineering for Futron Corporation, a Bethesda, Maryland based company providing risk management and aerospace safety and dependability services to government and commercial organizations including the Federal Aviation Administration, Department of Defense, NASA, Department of Energy, Westinghouse, Allied Signal and others.

Overall Colonel O'Connor has flown over 5000 hours in over forty types of aircraft, and has accumulated over 386 hours in space covering five and three quarter million miles in 253 orbits of the earth. His achievements include Naval Safety School Top Graduate, Naval Test Pilot School Distinguished Graduate Award, two Defense Superior Service Medals, the Distinguished Flying Cross, Navy Meritorious Service Medal, NASA Distinguished Service Medal, two NASA Outstanding Leadership Medals, two NASA Exceptional Service Medals, NASA Exceptional Achievement Medal, NASA Silver Snoopy Award, AIAA System Effectiveness and Safety Award, AIAA Barry M. Goldwater Education Award, and being named the *Aviation Week & Space Technology* Laureate (Space and Missiles).

ROBERT F. OVERMYER
Columbia and Challenger

Colonel Robert Franklyn Overmyer (July 14, 1936 - March 22, 1996) was a Marine Corps test pilot and USAF and NASA astronaut.

Overmyer was born in Lorain, Ohio, but he considered Westlake, Ohio to be his hometown. He graduated from Westlake High School in 1954, and later earned a Bachelor of Science degree in physics from Baldwin Wallace College in 1958, and a Master of Science degree in Aeronautics with a major in Aeronautical Engineering from the U.S. Naval Postgraduate School in 1964.

Overmyer entered active duty with the Marine Corps in January of 1958, and after completing Navy flight training in Kingsville, Texas was assigned to Marine Attack Squadron 214 in November 1959. Overmyer was then assigned to the Naval Postgraduate School in 1962 to study aeronautical engineering, and upon completion of his graduate studies served one year with Marine Maintenance Squadron 17 in Iwakuni, Japan. He was then assigned to the Air Force Test Pilots School at Edwards Air Force Base in California, and while there was chosen as an astronaut for the USAF Manned Orbiting Laboratory (MOL) Program in 1966.

The MOL program was cancelled in 1969, and soon afterward Overmyer was selected as a NASA astronaut. His first assignment with NASA was engineering development duties on the Skylab Program from 1969 until November

1971, and from November 1971 until December 1972 he was a support crew member for *Apollo 17* and was the launch capsule communicator. From January 1973 until July 1975 he was a support crew member for the Apollo-Soyuz Test Project, and was the NASA capsule communicator in the mission control center in Moscow. In 1976 he was assigned duties on the Space Shuttle Approach and Landing Test Program and was the prime T-38 chase pilot for Orbiter Free-Flights 1 and 3. Then in 1979 Colonel Overmyer was assigned as the Deputy Vehicle Manager in charge of finishing the manufacturing and tiling of *Columbia* at the Kennedy Space Center preparing it for its first flight. This assignment lasted until *Columbia* was transported to the launch pad in 1980.

Colonel Overmyer was the pilot for STS-5, the first fully operational flight of the Shuttle Transportation System, which launched from Kennedy Space Center on November 11, 1982. He was accompanied by spacecraft commander Vance D. Brand, who is also a Marine, and two mission specialists. STS-5, the first mission with a four-man crew, clearly demonstrated the space shuttle as fully operational by the successful first deployment of two commercial communications satellites from the orbiter's payload bay. The mission marked the first use of the Payload Assist Module, and its new ejection system. Numerous flight tests were performed throughout the mission to document shuttle performance during launch, boost, orbit, atmospheric entry and landing phases, and STS-5 was the last flight to carry the Development Flight Instrumentation package to support flight testing. A Getaway Special, three Student Involvement Projects and medical experiments were included on the mission, and the crew successfully concluded the five-day orbital flight of *Columbia* with the first entry and landing through a cloud deck to a hard-surface runway at Edwards Air Force Base in California.

Colonel Overmyer was the commander of STS-51-B, the Spacelab-3 mission, during which he commanded a crew of four astronauts and two payload specialists conducting a broad range of scientific experiments from space physics to the suitability of animal holding facilities. Mission 51-B was also the first shuttle flight to launch a small payload from the "Getaway Special" canisters. *Challenger* launched on April 29, 1985 from Kennedy Space Center, and landed at Edwards Air Force Base on May 6, 1985 after completing 110 orbits of the earth at an altitude of 190 nautical miles.

Colonel Overmyer retired from NASA and the Marine Corps in May 1986 after logging over 7,500 flight hours, with over 6,000 in jet aircraft. He was awarded the Distinguished Flying Cross, USAF Meritorious Service Medal for duties with the USAF Manned Orbiting Laboratory Program, the Meritorious Service Medal for duties as the Chief Chase Pilot and support crewman for the Shuttle Approach and Landing Test Program, an Honorary Doctor of Philosophy degree from Baldwin Wallace College, and the U.S. Naval Postgraduate School Distinguished Engineers Award.

Colonel Robert Overmyer died on March 22, 1996 in the crash of a Cirrus VK-30, a light aircraft he was testing, and is survived by his wife, Katherine, and three children.

ROBERT C. SPRINGER
Atlantis

Robert Clyde "Bob" Springer was a Marine aviator and an astronaut during the early years of NASA's Space Shuttle program.

Springer was born on May 21, 1942 in St. Louis, Missouri and was active in the Boy Scouts of America where he achieved its second highest rank, Life Scout. He graduated from Ashland High School in Ashland, Ohio, and received a commission in the Marine Corps following his 1964 graduation from the U.S. Naval Academy. Springer then attended the Marine Corps Basic School in Quantico, Virginia before reporting to the United States Navy's Air Training Command for flight training at Pensacola, Florida and Beeville, Texas.

Upon receiving his aviator wings in August of 1966 he was assigned to VMFA-513 at Marine Corps Air Station Cherry Point, North Carolina where he flew F-4 Phantom II fighters. He was subsequently assigned to VMFA-115 at Chu Lai in South Vietnam, where he completed three hundred F-4 combat missions. In June of 1968 Springer also served as an advisor to the South Korean Marine Corps in Vietnam, and flew 250 combat missions in O-1 Bird Dogs and UH-1 Iroquois "Huey" helicopters.

Springer returned to the United States to attend the U.S. Naval Postgraduate School in Monterey, California, and in

March 1971 was assigned to the Third Marine Aircraft Wing at El Toro where he became wing operations analysis officer. He flew UH-1Es in 1972 while with HML-267 at Camp Pendleton, and then went to Okinawa, Japan to fly Hueys with HML-367 of the 1st Marine Aircraft Wing. Springer then flew F-4 Phantom II fighters as an aircraft maintenance officer with VMFA-451 in Beaufort, South Carolina, and also attended Navy Fighter Weapons School.

A 1975 graduate of the U. S. Navy Test Pilot School at Patuxent River, Maryland, he served as Head of the Ordnance Systems branch and as a test pilot for more than twenty different types of fixed and rotary-wing aircraft, and in this capacity performed the first flights in the AHIT helicopter. He graduated from the Armed Forces Staff College in Norfolk, Virginia in 1978 and was assigned to Headquarters Fleet Marine Force, Atlantic, where he assumed responsibility for joint operational planning for Marine Forces in NATO and the Middle East. He was serving as aide-de-camp for the Commanding General, Fleet Marine Force, Atlantic when advised of his selection by NASA in May of 1980.

Springer became an astronaut in August of 1981, and his technical assignments included support crew for STS-3, concept development studies for the Space Operations Center, and the coordination of various aspects of the final development of the Remote Manipulator System for operational use. He also worked at Mission Control in the Lyndon B. Johnson Space Center as the CAPCOM for seven flights between 1984 and 1985.

Springer was responsible for astronaut office coordination of Design Requirements Reviews and Design Certification Reviews. These review efforts encompassed the total recertification and reverification of the NSTS prior to STS-

26's return to flight status. He flew as a mission specialist aboard STS-29 in 1989, and on STS-38 in 1990.

STS-29 aboard Space Shuttle *Discovery* was launched in March of 1989, and during eighty orbits of the earth on this highly successful five-day mission the crew deployed a Tracking and Data Relay Satellite and performed numerous secondary experiments, including a Space Station "heat pipe" radiator experiment, two student experiments, a protein crystal growth experiment, and a chromosome and plant cell division experiment. In addition, the crew took over four thousand photographs of the earth using several types of cameras, including the IMAX 70 mm movie camera. Mission duration was 119 hours, and concluded with a landing at Edwards Air Force Base.

Space Shuttle *Atlantis* was launched at night from Kennedy Space Center in November of 1990, and during the five-day mission the crew conducted Department of Defense operations. After eighty orbits of the earth *Atlantis* and her crew landed back at the Kennedy Space Center in the first Shuttle recovery in Florida since 1985.

Springer retired from NASA and the Marine Corps in December of 1990, and over the course of his career logged over 237 hours in space and 4,500 hours flying time, including 3,500 hours in jet aircraft. He was awarded the Navy Distinguished Flying Cross, Bronze Star, Twenty-one Air Medals, two Navy Commendation Medals, the Navy Achievement Medal, NASA Space Flight Medal, Combat Action Ribbon, Presidential Unit Citation, Navy Unit Citation, and various Vietnam campaign ribbons and service awards.

FREDERICK W. STURCKOW
Endeavour, Discovery and Atlantis

Frederick Wilford "CJ" Sturckow is a Marine Corps officer and NASA astronaut.

Sturckow was born La Mesa, California on August 11, 1961, but considers Lakeside, California to be his hometown. He graduated from Grossmont High School in La Mesa, California, in 1978, and received a Bachelor of Science degree in mechanical engineering from California Polytechnic State University in 1984.

Sturckow was commissioned in December of 1984, was an Honor Graduate of The Basic School, and earned his wings in April of 1987. Following initial F/A-18 training at VFA-125 he reported to VMFA-333, MCAS Beaufort, South Carolina. While assigned to VMFA-333 he made an overseas deployment to Japan, Korea, and the Philippines and was then selected to attend the Navy Fighter Weapons School (Top Gun) in March of 1990. In August 1990 he deployed to Sheik Isa Air Base, Bahrain for a period of eight months, and flew a total of forty-one combat missions during Operation Desert Storm. In January 1992 he attended the United States Air Force Test Pilot School at Edwards AFB, California, and in 1993 reported to the Naval Air Warfare Center, Aircraft Division, Patuxent River, Maryland for duty as the F/A-18 E/F Project Pilot. Sturckow also flew a wide variety of projects and classified programs as an F/A-18 test pilot.

Selected by NASA in December 1994, Sturckow reported to the Johnson Space Center in March of 1995, completed a year of training and evaluation, and was assigned to work technical issues for the Vehicle Systems and Operations Branch of the Astronaut Office. He currently serves as Deputy for the Shuttle Operations Branch of the Astronaut Office, and also serves as Lead for Kennedy Space Center Operations Support.

Before commanding the STS-128 mission launched on August 28, 2009 Sturckow was a veteran of three space flights aboard shuttles *Endeavour, Discovery and Atlantis.* He served as pilot on STS-88 in 1998, which was the first International Space Station assembly mission, piloted STS-105 in 2000, and was also the commander of STS-117.

Sturckow's nickname "CJ" stands for "Caustic Junior." It was given to him when he was a young Marine because he resembled a squadron commander who was appropriately called "Caustic." He has logged over four thousand flight hours, flown over fifty different aircraft, and has logged over 904 hours in space. His awards include the Defense Superior Service Medal, Single Mission Air Medal with Combat "V," and Navy and Marine Corps Commendation Medal.

TERRENCE W. WILCUTT
Atlantis and Endeavour

Colonel Terrence Wade Wilcutt is a Marine Corps officer and NASA astronaut who is a veteran of four shuttle missions and is currently the Deputy Director, Safety and Mission Assurance, at Johnson Space Center.

Born on 31 October 1949 and raised in Louisville, Kentucky, Wilcutt earned a degree in mathematics from Western Kentucky University. He then taught high school math for two years before entering the Marine Corps where he trained as a pilot and flew the F/A-18 before being assigned to the Naval Aircraft Test Center and working on classified aircraft programs.

Wilcutt was commissioned in the Marine Corps in 1976 and earned his wings in 1978. Following initial F-4 Phantom training with VMFAT-101 he reported to VMFA-235 in Kaneohe Bay, Hawaii. While assigned to VMFA-235, Wilcutt attended the Naval Fighter Weapons School (Top Gun) and made two overseas deployments to Japan, Korea, and the Philippines. In 1983, he was selected for F/A-18 conversion training, and served as an F/A-18 Fighter Weapons and Air Combat Maneuvering Instructor for VFA-125 in Lemoore, California. In 1986 Wilcutt was selected to attend the United States Naval Test Pilot School, and following graduation from USNTPS he was assigned as a test pilot/project officer for Strike Aircraft Test Directorate at the Naval Aircraft Test Center in Patuxent River, Maryland. While assigned to SATD Wilcutt flew the F/A-18 Hornet, the A-7 Corsair II, the F-4 Phantom, and various other aircraft while serving as a test pilot/project officer in a wide variety of projects and classified programs.

Wilcutt was selected as an astronaut candidate in 1990 and piloted missions STS-68 in 1994 and STS-79 in 1996. He then commanded mission STS-89 to the Mir space station in 1998, and STS-106 to the International Space Station in 2000.

STS-68 *Endeavour* was part of NASA's Mission to Planet Earth, and its Space Radar Lab-2 was the second flight of three advanced radars called SIR-C/X-SAR (Spaceborne Imaging Radar-C/X-Band Synthetic Aperture Radar) and a carbon-monoxide pollution sensor, MAPS (Measurement of Air Pollution from Satellites). SIR-C/X-SAR and MAPS operated together in *Endeavour's* cargo bay to study the Earth's surface and atmosphere, creating radar images of Earth's surface environment and mapping global production and transport of carbon monoxide pollution. Real-time crew observations of environmental conditions, along with over 14,000 photographs, aided the science team in interpreting the SRL data. The SRL-2 mission was a highly successful test of technology intended for long-term environmental and geological monitoring of planet Earth. Mission duration was 11 days, 5 hours, and 46 minutes, with *Endeavour* traveling 4.7 million miles in 183 orbits of the Earth.

STS-79 *Atlantis*, the fourth in the joint American-Russian Shuttle-Mir series of missions, launched from and returned to land at Kennedy Space Center in September of 1996. *Atlantis* rendezvoused with the Russian MIR space station and ferried supplies, personnel, and scientific equipment to this base 240 miles above the Earth. The crew transferred over 3.5 tons of supplies to and from Mir, and exchanged U.S. astronauts on Mir for the first time - leaving John Blaha, and bringing Shannon Lucid home after her record six month stay.

STS-89 in 1998 was the eighth Shuttle-Mir docking mission during which the crew transferred more than 9,000 pounds of scientific equipment, logistical hardware and water from Space Shuttle *Endeavour* to Mir. In the fifth and last exchange of a U.S. astronaut, STS-89 delivered Andy Thomas to Mir and returned with David Wolf.

STS-106 *Atlantis* in September of 2000 was a twelve-day mission during which the crew successfully prepared the International Space Station for the arrival of the first permanent crew. The five astronauts and two cosmonauts delivered more than 6,600 pounds of supplies and installed batteries, power converters, life support, and exercise equipment on the space station. Two crew members performed a space walk in order to connect power, data and communications cables to the newly arrived Zvezda Service Module and the space station itself. *Atlantis* orbited the Earth 185 times, and covered 4.9 million miles in 11 days, 19 hours, and 10 minutes.

Colonel Wilcutt amassed over 6,600 flight hours in more than thirty different aircraft, and has been awarded the NASA Outstanding Leadership Medal, NASA Distinguished Service Medal, NASA Exceptional Service Medal, Distinguished Flying Cross, Defense Superior Service Medal, Defense Meritorious Service Medal, and the Navy Commendation Medal.

GEORGE D. ZAMKA
Discovery

George David "Zambo" Zamka is a NASA astronaut and Marine Corps aviator who piloted the Space Shuttle *Discovery* in its October 2007 mission to the International Space Station and served as the commander of mission STS-130 in February of 2010.

Zamka was born in Jersey City, New Jersey in 1962 and was raised in New York City, Irvington, New York, Rochester Hills, Michigan, and his mother's hometown of Medellín, Colombia. He graduated from Rochester Adams High School in Michigan in 1980.

Zamka graduated with a Bachelor of Science degree in Mathematics from the United States Naval Academy in 1984 and was commissioned as a Second Lieutenant in the Marine Corps. He received A-6E Intruder training at Naval Air Station Whidbey Island, Washington from 1985 to 1987, and was then assigned to VMA(AW)-242 at Marine Corps Air Station El Toro in California. In addition to flight safety and administration he was a Squadron Weapons and Tactics instructor, and in 1990 he trained as an F/A-18D Hornet pilot and was assigned to VMFA(AW)-121, with whom he flew sixty-six combat missions during Operation Desert Storm.

In 1993 he was assigned to the 1st Battalion, 5th Marines at Marine Corps Base Camp Pendleton, California as a forward air controller. Then in December of 1994 he graduated from the U.S. Air Force Test Pilot School, following which he served as an F/A-18 Hornet test pilot and operations officer.

In 1997 Zamka earned a Master of Science degree in Engineering Management from the Florida Institute of Technology, and in 1998 he returned to VMFA(AW)-121 and deployed to MCAS Iwakuni, Japan.

In June of 1998 Zamka was selected for the NASA astronaut program, and he reported for training in August. He served as lead for the shuttle training and procedures division and as supervisor for the astronaut candidate class of 2004. Zamka made his first spaceflight as the pilot of mission STS-120, and subsequently commanded STS-130.

Zamka has over 3500 flight hours in more than thirty different aircraft, and has been awarded six Navy Strike Air Medals, the Navy Commendation Medal with Combat V, and various other military service and campaign awards. He was also a Distinguished Graduate of United States Naval Academy.

PHOTO GALLERY

The future Governor of South Dakota, Medal of Honor recipient Captain Joe Foss. He was much more than a fighter pilot!

Senator Paul Douglas (top left) enlisted in the Marine Corps during WWII as a private at the age of fifty. The Visitors' Center at MCRD Parris Island is named in his honor. Senator Zell Miller of Georgia (top right) was a sergeant in the Corps.

Colonel William Eddy with King Abdul-Aziz Al Saud, and in France during World War I. He is third from the right.

Story Musgrave as a young Marine, and as a weightless astronaut. Below, Colonel Cabana can be excused for the Navy sweatshirt because he attended the Academy, and his interest in beating Army.

Senator John Warner with wife Elizabeth Taylor, and in the Oval Office watching the President sign a piece of legislation which he sponsored. That's General Peter Pace on the right, who was the first Marine to serve as Chairman of the Joint Chiefs of Staff.

Captain and future Senator Charles Robb marries the daughter of President Lyndon B. Johnson in the East Room of the White House.

Governor Sid McMath of Arkansas (top left), who served on Guadalcanal and Bougainville during WWII, and Senator Jim Webb (top right), who was awarded the Navy Cross in Vietnam. Below, Eugene Stoner and Mikhail Kalashnikov, inventor of the AK-47.

Wake Island, where Major (and future Congressman) James Devereaux and his Marines valiantly fought the Japanese, from the air and on the ground.

Secretary of State Condoleezza Rice appointed former Commandant James Jones as a special envoy for Middle East security prior to his becoming National Security Advisor in the Obama Administration.

Maryland Senator Daniel Brewster, seen here in the Oval Office with President Lyndon Johnson, fought on Guam and Okinawa and was wounded seven times.

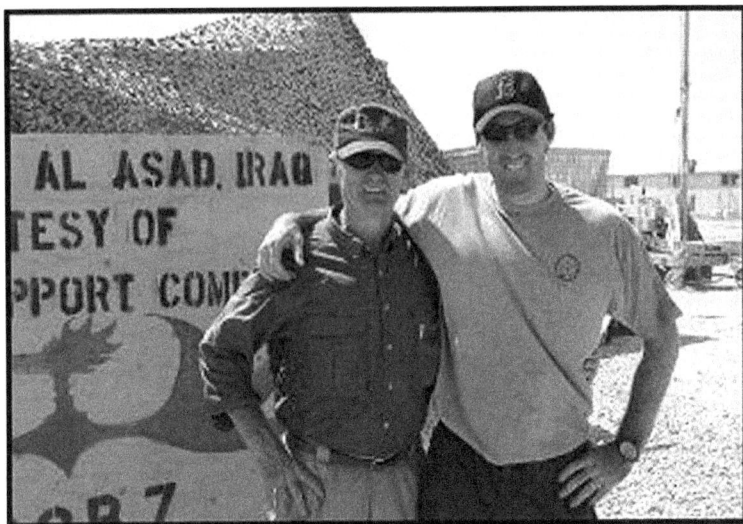

Bing West with his son and fellow Marine Owen in Al Asad, Iraq.

Fred Haise, Jim Lovell and Jack Swigert aboard *USS Iwo Jima* after returning from their aborted Apollo 13 mission.

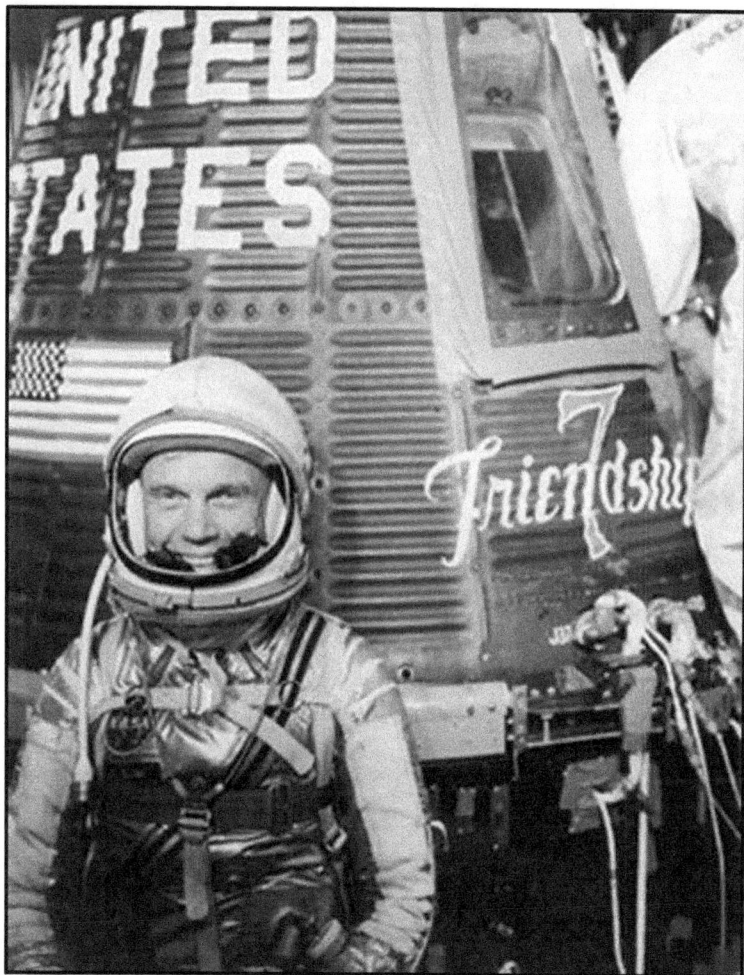

Colonel and future Senator John Glenn poses with his Mercury capsule, *Friendship 7*, before embarking on his historic flight.

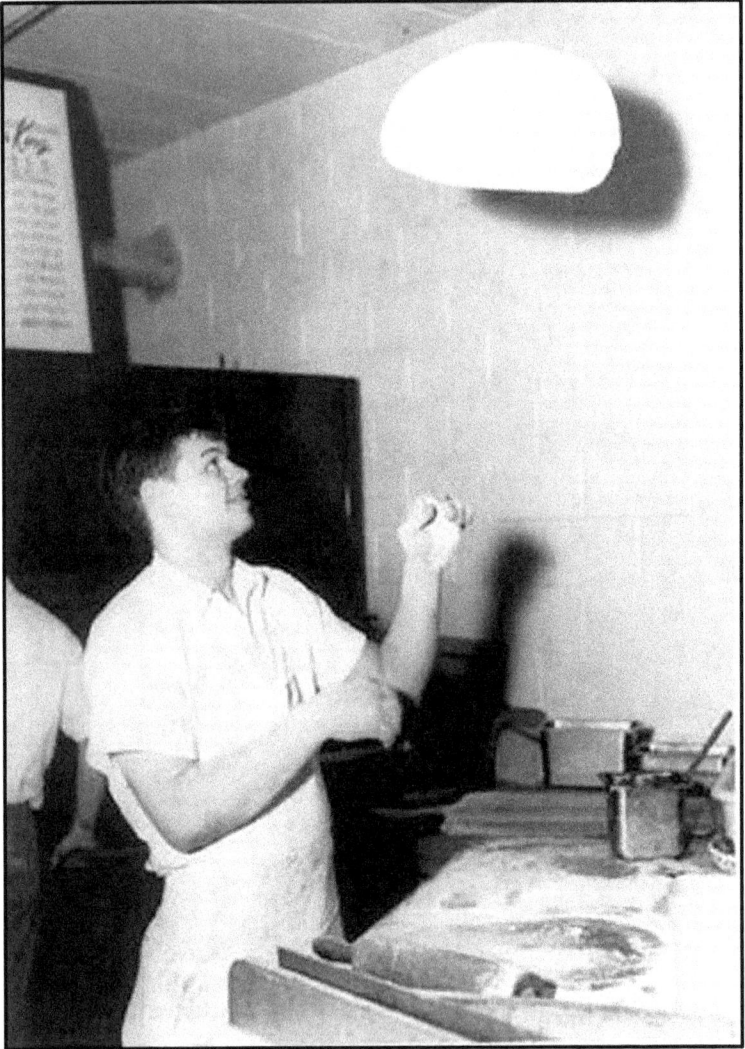

Tom Monaghan making a pizza during the 1960s. He probably never imagined his business would someday grow to become Domino's.

Above, Congressman Duncan Hunter during one of his tours of duty as a Marine Major in Iraq, and at left, the legendary Joe Foss poses with an F4F Wildcat, the type of aircraft he flew on his way to becoming an Ace and earning the Medal of Honor at Guadalcanal.

"If you want to visit his grave, don't look for him near the Kennedy Eternal Flame, where so many politicians are laid to rest. Look for a small, common marker shared by the majority of our heroes. Look for the marker that says, 'Michael J. Mansfield, PFC, U.S. Marine Corps.'" – Colonel James M. Lowe